T0214078

SpringerBriefs in Earth Sciences

SpringerBriefs in Earth Sciences present concise summaries of cutting-edge research and practical applications in all research areas across earth sciences. It publishes peer-reviewed monographs under the editorial supervision of an international advisory board with the aim to publish 8 to 12 weeks after acceptance. Featuring compact volumes of 50 to 125 pages (approx. 20,000–70,000 words), the series covers a range of content from professional to academic such as:

- timely reports of state-of-the art analytical techniques
- bridges between new research results
- snapshots of hot and/or emerging topics
- literature reviews
- in-depth case studies

Briefs will be published as part of Springer's eBook collection, with millions of users worldwide. In addition, Briefs will be available for individual print and electronic purchase. Briefs are characterized by fast, global electronic dissemination, standard publishing contracts, easy-to-use manuscript preparation and formatting guidelines, and expedited production schedules.

Both solicited and unsolicited manuscripts are considered for publication in this series.

More information about this series at http://www.springer.com/series/8897

Sofia B. Shah

Heavy Metals
in Scleractinian Corals

 Springer

Sofia B. Shah
Department of Chemistry
Fiji National University
Lautoka, Fiji

ISSN 2191-5369   ISSN 2191-5377 (electronic)
SpringerBriefs in Earth Sciences
ISBN 978-3-030-73612-5   ISBN 978-3-030-73613-2 (eBook)
https://doi.org/10.1007/978-3-030-73613-2

This Springer imprint is published by the registered company Springer Nature Switzerland AG
The registered company address is: Gewerbestrasse 11, 6330 Cham, Switzerland

Preface

Increasing interest in ocean chemistry is prevalent nowadays owing to the introduction of pollutants into the ocean via land-based sources, ocean dumping and shipping activities. The majority of the pollutants originate from land-based sources and are transported into the ocean via river flows, submarine groundwater discharge and atmospheric deposition. These pollutants comprise sewage and nutrients (leading to eutrophication phenomena); persistent organic pollutants (POPS); oil spills and petroleum derivatives and heavy metals and organometallic compounds. Heavy metals are one such pollutant that is of growing interest due to their potential toxicity and the ability to bioaccumulate in the marine environment and therefore its effect on corals is deliberated here. Heavy metal pollution is recognized as a major environmental concern.

This monograph is designed to serve as an exceptional interdisciplinary resource material for upcoming marine researchers and/or chemists, post graduate students and those interested in marine pollution. It delves into the effects of heavy metal pollutants on *scleractinian* corals, which is part of research work undertaken for MSc thesis titled, "Study of Heavy Metal Accumulation in *Scleractinian corals* of Viti Levu, Fiji Islands". The monograph contributes to an overview of heavy metals, their sources to the marine environment, their effects on *scleractinian* corals and techniques for analysis. Furthermore, this monograph examines and evaluates the current state of knowledge on the status quo of heavy metals in the marine environment, responses of corals to heavy metals and heavy metal regulation in corals and tabulates analytical techniques for determining heavy metals and suggests possible recommendations for the future.

Heavy metals can occur naturally as well as anthropogenically and enter into the marine environment via direct and indirect sources. Heavy metals are regarded as perpetual additions to the marine environment and do not get broken down by bacterial action. They are considered toxic, persistent and non-biodegradable as they cannot be broken down by chemical or biological processes. Once in the marine environment, the heavy metals have a series of deleterious sub-lethal and lethal effects on marine biota. The toxicity of a particular heavy metal is closely associated to its chemical speciation.

Scleractinian corals are the architects of coral reefs and hence good indicators of environmental changes. Being sedentary in nature, *scleractinian* corals are a good representative of their environment. Heavy metals get incorporated into the coral skeleton through the substitution of dissolved metal species via calcium substitution and remain embedded forever since the new growth covers the old carbonate surface.

Heavy metal regulation in marine invertebrates can be obtained through excretion, impaired uptake, detoxification, storage, sequestration of metals, metal-binding to proteins such as glutathione and metallothioneins and release of metals via increased production of mucus or nematocyst discharge, and *scleractinian* corals are no exception to this. Various analytical techniques are used to analyze heavy metal concentrations in different components of diverse corals.

Lautoka, Fiji   Sofia B. Shah

Acknowledgements

Special notes of appreciation go to the Almighty and several individuals whose expertise made this monograph possible. Thanks to Professor Amanda Reichelt-Brushett of Southern Cross University for her immense support, guidance, mentoring and thorough review of the monograph. Thank you Dr. Matakite Maata of the University of the South Pacific for your support and review of the monograph and Dr. Todd Dennis of Fiji National University for your help with textbooks related to Coral Reefs. Heartfelt appreciation to my family for your understanding and patience. The help of these generous and talented people is greatly appreciated.

Contents

Chapter 1
Heavy Metals in the Marine Environment—An Overview

Abstract Heavy metals are the natural constituents of the Earth's crust, found in very low concentrations; however, human activities such as industrial waste discharges, agricultural practices, coastal construction and dredging have inevitably increased the metal concentrations in the marine environment. Heavy metals are classified as both essential and non-essential elements. Essential heavy metals have known biological roles and are only toxic above threshold concentrations whilst non-essential heavy metals lack any known biological role in marine invertebrates and exhibit high degree of toxicity if allowed to accumulate at metabolically active sites. Heavy metals are non-biodegradable, persistent and toxic to the environment, thus causing serious eco-toxicological problems. Heavy metals tend to bioaccumulate and the extent of their bioaccumulation is dependent on the total amount, the bioavailability of each metal in the environmental medium and the route of uptake, storage and excretion mechanisms. Metal speciation influences metal bioavailability and toxicity to biota, its transportation and mobilization, and its interaction with the environment. The actual metal speciation is influenced by factors such as pH, the types and concentrations of inorganic ligands and organic ligands and colloidal species present.

Keywords Heavy metals · Metal speciation · Bioaccumulation · Toxicity

1.1 Heavy Metals

Heavy metals are the natural components of the Earth's crust and regarded as the members of an ill-defined subset of elements exhibiting metallic properties, including transition metals, some metalloids, lanthanides and actinides. Lata and Rohindra (2002) and Shah (2008) reported that heavy metals include metals and metalloids, having atomic weights between 63.5 and 200.6 g/mol and density greater than 4.5 g/cm^3. Okoro et al. (2012) highlighted that there are over 50 elements classified as heavy metals, but only 17 considered both very toxic and relatively accessible.

Heavy metals are the natural constituents of the marine and freshwater environment, commonly found in very low concentrations (Ansari et al. 2004). Natural processes that bring about the occurrence of heavy metals in the environment include comets, erosion, volcanic eruptions, and the weathering of minerals; with these heavy metals characteristically present in insoluble forms such as in mineral structures, precipitated or complex forms (Ayangbenro and Babalola 2017).

Anthropogenic activities such as mine drainage, offshore oil and gas exploration, industrial (pesticides, paints, leather, textile, fertilizers, pharmaceuticals) and domestic effluents, wastewaters, agricultural runoff, acid rain waste discharges, and coastal construction and dredging have inevitably increased the metal concentrations in the marine environment (Ansari et al. 2004; Fu and Wang 2011). Heavy metals from anthropogenic sources typically have a high bioavailability due to their soluble and mobile reactive forms (Ayangbenro and Babalola 2017). Metal inputs are permanent additions to the marine environment and do not get broken-down by bacterial action and are rendered inert (Hudspith et al. 2017). However, they can only be converted into less toxic species (Ayangbenro and Babalola 2017).

The term 'heavy metal' is used synonymously with 'trace metal' (Rainbow 1995) and are considered as trace elements because of their presence in trace concentrations (ppb range to less than 10 ppm) in various environmental matrices (Kabata-Pendia 2001). Hooda (2010) also mentioned that 'heavy metal' is the most popularly used and widely recognized term. Hence, this monograph adopted the use of heavy metals in accordance with the definitions and descriptions provided by Rainbow (1995), Kabata-Pendia (2001) and Hooda (2010).

Heavy metal comprises of both essential and non-essential trace metals (Rainbow 1995; Richir and Gobert 2016). Essential elements recognized by WHO (1996) include copper [Cu], selenium [Se], chromium [Cr], molybdenum [Mo], zinc [Zn] and iron [Fe]. Essential elements are vital components of enzymes, respiratory proteins and certain structural elements of organisms (Brown and Depledge 1985), but above certain threshold concentrations, they are potentially toxic to both humans and aquatic biota. Being essential micronutrients, trace metals are also important in controlling marine productivity (Chen et al. 2015).

Non-essential elements consist of cadmium [Cd], lead [Pb], or mercury [Hg]; and play no physiological role and are often toxic even in small quantity (Nordberg et al. 2007). Since non-essential elements lack any known biological role in marine invertebrates, they would exhibit high degree of toxicity if allowed to accumulate at metabolically active sites. These heavy metals may imitate essential metals and thus gain access to important molecular targets (Chiarelli and Roccheri 2014).

Heavy metals are serious pollutants because of toxicity, persistence, ubiquitous nature and non-biodegradability in the environment (Pekey 2006; DeForest et al. 2007; Naser 2013; Wu et al. 2016) and have a series of deleterious sub-lethal and lethal effects on marine biota (Peters et al. 1997). Heavy metals tend to bioaccumulate (DeForest et al. 2007; Naser 2013), meaning that their concentration increases in a biological organism over time. The extent of bioaccumulation of metals is dependent on the total amount, the bioavailability of each metal in the environmental medium and the route of uptake, storage and excretion mechanisms (Chiarelli and Roccheri 2014).

The transportation of heavy metals to long distances is via atmospheric and hydrological processes (Guzmán and Jiménez 1992; Marx and McGowan 2010). McLusky et al. (1986); Rainbow (1997) and Marsden and Rainbow (2004) reported that heavy metals occur in a variety of forms and species which affects their toxicity in the environment. Heavy metals present a serious threat to the marine environment due to their potential toxic effects (Pastorok and Bilyard 1985; Negri and Heyward 2001).

Once released into the aquatic environment, trace elements are transported and distributed between the aqueous phase and sediment through the process of adsorption and desorption (Mondal et al. 2017). The trace elements can occur in several forms such as a free ion; as an ion bound to different ligands (formation of complex compounds); as the precipitated molecule of a compound, suspended in the liquid phase or adsorbed on the surface of suspended or colloidal matter (e.g. Manahan 2003). Ionic species are more bioavailable and toxic compared to those bound to particles and organic compounds (Kroon et al. 2020). A significant quantity of free metal ions forms complexes via adsorption, hydrolysis, and co-precipitation and gets deposited in the sediment while only a small portion of ions remain dissolved in the water column (Bastami et al. 2015).

Heavy metals have greater detrimental environmental effects when present in a water body that has a low pH because changes in metal speciation result in a shift to more toxic ionic metal species (Bielmyer-Fraser et al. 2018). Furthermore, heavy metal contamination may have devastating effects on the ecological balance of the recipient environment and a diversity of aquatic organisms (Ashraf 2005; Vosyliene and Jankaite 2006; Farombi et al. 2007).

1.2 Toxicity of Heavy Metals

All heavy metals can be toxic if present above threshold concentrations. Toxicity of the metals is the capability of a metal to cause undesirable effects on organisms (Ayangbenro and Babalola 2017). Ansari et al. (2004) reported that metal toxicity depends on factors like chemical speciation of metals in the aquatic environment, presence of other toxicants, environmental conditions including temperature, pH, salinity and dissolved oxygen; condition of the organism tested; kinetics of toxic reactions and adaptation of the organism to the absorption of metals. Furthermore, factors such as the type of metals, their biological roles, and the type of organisms exposed to it also affect metal toxicity (Akan et al. 2010). In addition to this, the toxicity of the metal is also dependent on the amount accessible to organisms, the absorbed dosage, the route and duration of exposure (Mani and Kumar 2014).

Elevated levels of heavy metals in the marine environment could lead to sub-lethal effects in the marine organism including changes in physiology, tissues, biochemistry, behavior and reproduction. Chiarelli and Roccheri (2014) reported that often metals penetrate the cells via transport mechanisms normally used for other purposes and are irreversibly accumulated in cells where they interact with cellular components and molecular targets. Kim et al. (2014) and Tamás et al. (2014) mentioned that the toxicity of the metals has been associated with blockage of oxidative phosphorylation, glutathione depletion and inhibition of antioxidant enzymatic activity, production of ROS (reactive oxygen species), DNA damage and inhibition of relevant repair mechanisms, and protein misfolding disorders.

Heavy metals produce adverse effects on the health of humans and other living beings in terrestrial and aquatic environment and affect the food chain (Das et al. 2013). Even though most marine organisms tend to accumulate heavy metals from the environment, they are also capable of storing, removing or detoxifying many of them but these capabilities tend to differ between species, as some species could be more tolerant than others (De Pooter 2013).

Metal related diseases have been identified in countries undergoing rapid industrialization (Esslemont 1998) and these are correlated to environmental disasters, some of which are mentioned below:

- Late 1800s: Cr toxicity in Scotland (Esslemont 1998);
- 1932-Sewage containing Hg released in Minamata Bay, Japan. There was Hg accumulation in the sea creatures, leading to Hg poisoning in the population (Harada 1995; Lenntech 2006);
- 1910–1940s: waste sludge containing Cd from the Kamioka zinc mines in the River Jinzu catchment of Japan gradually contaminated rice paddies and drinking water in the area. Cd uptake by rice enhanced by the low pH of the contaminated paddy soils. Ingestion of contaminated rice over many years caused kidney dysfunction and osteomalacia leading to fragile and deformed bones (Rieuwerts 2015); and this led to the disease known as itai itai disease. Itai itai was referred to as such because of the excruciating pain experienced by the people who had consumed contaminated rice;
- 1950s: Cd toxicity in Japan (Esslemont 1998);
- 1952: mass mortality in Japan of human population due to the consumption of Hg polluted fish-referred to as Minamata Syndrome (Lenntech 2006);
- 1986: water used to extinguish a major fire carries ~30t of fungicide containing Hg into the Upper Rhine, at Sandoz in Germany, thereby killing fish (Giga 2009);
- 1998: toxic chemicals from a burst dam contaminate Spanish nature reserve (Lenntech 2006).
- 2010: fish kill resulting from Cu pollution in China: a chemical spill at a Cu mining-smelting complex in southeast china entered a major river, resulting in a significant pollution event (Rieuwerts 2015).
- 2011: Cr contamination of drinking water in China: hexavalent Cr contamination of public water supplies in Yunnan province due to several thousand tonnes of waste tailings from the production of the tanning chemical chromium sulphate, dumped illegally on wasteland (Rieuwerts 2015).
- 2014: As pollution in the USA, where 8×10^4 ton of coal spilled into the Dan River in North Carolina, USA contaminating the river with As and other toxic elements, including Cr (Rieuwerts 2015).

1.3   Common Heavy Metals in the Marine Environment

Heavy metals are eminent marine pollutants; naturally occurring as well as anthropogenically introduced into the marine environment. Continental runoff and atmospheric deposition are the primary natural inputs of trace elements in the marine environments while the major anthropogenic sources include mining and smelting (Richir and Gobert 2016).

Al-Rousan et al. (2007) further elaborates that natural and anthropogenic sources of heavy metals in the marine environment include terrigenous input during flash floods that transport terrestrial material, ground water input, agricultural activities, increase in land traffic, sedimentation caused by filling and coastal construction and dredging, oil spills and discharges, industrial discharges (fertilizers, plastic stabilizers), ship-based sewage and solid waste, soft waste dumping (alloys, dyes, automobile tyres, antifouling paints and galvanizing materials), and shipment of mineral products (mainly phosphates).

The transportation of heavy metals to the marine environment is through natural and anthropogenic induced process as dissolved species in water or in association with suspended sediments probably affecting the primary productivity and coral growth (Al-Rousan 2012). Heavy metals also enter the marine environment through atmospheric deposition, which is also regarded as an important exposure pathway.

Sedimentation contributes significant amount of heavy metals to the marine environment especially during heavy rainfall resulting in flash flooding and dust storms. The sediments can be of terrestrial origin such as run-off from logging catchments, or coastal development, or from poor farming practices or of riverine inputs. Sediments are composed of oxides of Fe and Mn (manganese), minerals, terragenic materials and organic matter (Rodrigo 1989). In marine environments adjacent to arid landmasses where fringing reefs usually occur, aeolian transport of fine particles can be the primary source of heavy metals (Esslemont 1998). Once in the water column, heavy metals are associated with sediment as well as particulate organic matter (Bastidas et al. 1999). Heavy metals can contribute to degradation of marine ecosystems by reducing species diversity and abundance and through accumulation of metals in living organisms and food chains (Hosono et al. 2011; Naser 2013).

Described below are some common heavy metals present in the marine environment. The major sources of these metals to the marine environment; roles of these metals in aquatic environment (if any) and their bioaccumulation and biomagnification properties are deliberated here.

First, a brief note on the definitions of bioaccumulation and biomagnification as used in the context of this monograph. Bioaccumulation is a good integrative indicator of chemical exposures of organisms in polluted ecosystems (Philips and Rainbow, 1994).

Bioaccumulation is the process of the intake of a chemical and its concentration in the organism by all potential means such as contact, respiration and ingestion (Alexander 1999) relative to that in the environment. The uptake of the chemical occurs either directly from exposure to a contaminated medium or by consumption of food containing the chemical (Mondal et al. 2018). Metal bioaccumulation can be complex and is influenced by multiple routes of exposure and geochemical effects on bioavailability (Luoma and Rainbow 2005). Bioavailability on the other hand, refers to the proportion of an element available for uptake from the environment within a time frame. Biomagnification occurs when the chemical is passed up the food chain to higher trophic levels, such that in predators it exceeds the concentration to be expected where equilibrium prevails between an organism and its environment (Neely 1980; Alexander 1999).

1.3.1 Cadmium (Cd)

Cd mostly occurs in sulfide minerals, which also contains Zn, Pb or Cu. Cd was first used in World War I as a substitute for tin and in paint industries as a pigment (Jaishankar et al. 2014). It is also used in electroplating, batteries, paint dyes, stabilizers for PVC (poly vinyl chloride) and alloys (Wuana and Okieimen 2011; Jaishankar et al. 2014).

Cd is a highly toxic environmental pollutant (Ansari et al. 2004; Bat et al. 2012; Chiarelli and Roccheri 2014; Wang et al. 2014a); and a non-essential metal as well (Chiarelli and Roccheri 2014). It mostly enters the marine environment via atmospheric loading, riverine discharges and agriculture (elevated levels in most superphosphate fertilizers) (Wuana and Okieimen 2011); and through the use of detergents and refined petroleum products (Gautam et al. 2014). It is noteworthy to mention that Cd is largely associated with Zn extraction and phosphate fertilizers (Loganathan and Hedley 1997; Howe et al. 2014). Furthermore, Reichelt-Brushett (2012) reported that Cd is of particular concern for marine environments that receive runoff from agricultural and mining operations.

Ansari et al. (2004) and Rieuwerts (2015) reported that surface seawater Cd acts as a micronutrient in marine phytoplankton, facilitating photosynthesis. Cd interferes with the metallothioneins (metal-binding proteins) activity and hence disrupts the homeostasis in an organism's body (Klaassen et al. 2009).

Cd is very bio persistent but has few toxicological properties and, once absorbed by an organism, remains resident for many years (Wuana and Okieimen 2011) and it does not have biomagnifying properties.

1.3.2 Copper (Cu)

Cu occurs naturally in rocks, soil, water, sediment and in air with generally low concentrations. Cu is used in the production of alloys (such as bronze, brass and copper-nickel alloy), electrical wiring, plumbing, roofing, machinery and electronics (Rieuwerts 2015). Cu mostly enters the marine environment via activities such as mining, smelting, discharge of wastewater, the use of wood preservatives for coastal waterfront structures (Brown and Eaton 2001), dumping of sewage sludge (Pastorok and Bilyard 1985; Guzmán and Jiménez 1992), using additives in drinking water to control algal growth (Rieuwerts 2015), agriculture (the use of copper-based fungicides and as impurities in many pesticides, fungicides and algaecides), ship groundings resulting in localized contamination (Haynes et al. 2002; Negri et al. 2002; Smith et al. 2003) and intensive animal farming.

Cu is also a major component of antifouling paints (Claisse and Alzieu 1993; Howe et al. 2014), present in industrial, urban and agricultural discharges (Mance 1987; Reichelt-Brushett and Harrison 1999) and is a component of some fungicides and herbicides that are used on coastal agricultural crops (Cremlyn 1979).

Smith et al. (2003) mentioned that tributyltin (TBT) had been applied on large vessels together with Cu and Zn as an antifouling biocide. The biocides are specifically designed and chosen to prevent settlement and growth of algae and invertebrates (Jones 2007) on marine vessels. TBT rapidly kills organisms such as mussels and barnacles, which attach themselves on ship hulls and hard surfaces. TBT was the most widely used active component in antifouling paints; however, a global ban was sanctioned in 2008 on the application of TBT paints (Tornero and Hanke 2016). TBT caused negative environmental impacts signs of imposex in marine invertebrate species and the deformation of oyster shells. Hence, biocides such as copper (I) salts, mainly in the form of copper oxide (Cu_2O) and copper thiocyanate (CuCHNS), have been the main alternatives to TBT in many antifouling coatings (Tornero and Hanke 2016).

It is well documented that Cu is an essential element required by all living organisms; however, at higher concentrations it may accumulate and cause toxicity in marine organisms (Bielmyer et al. 2005, 2010, 2012; Kim et al. 2008; Main et al. 2010; Bielmyer and Grosell 2011; Patel and Bielmyer-Fraser 2015; Siddiqui and Bielmyer-Fraser 2015; Siddiqui et al. 2015; Bielmyer-Fraser et al. 2018). The speciation and bioavailability of Cu in seawater is greatly reliant on seawater chemistry (Millero et al. 2009; Zeng et al. 2015).

Cu mostly exerts toxicity by altering enzyme function, causing oxidative stress, disrupting ionoregulation, and/or disrupting acid/base balance in aquatic organisms (Crespo and Karnaky 1983; McGeer et al. 2000; Bielmyer et al. 2005; Grosell 2011; Patel and Bielmyer-Fraser 2015; Siddiqui and Bielmyer-Fraser 2015; Siddiqui et al. 2015; Bielmyer-Fraser et al. 2018). Cu causes endocrine disruption in aquatic organisms (Rieuwerts 2015).

Solomon (2009) reported that Cu is one of the most toxic metals to aquatic organisms and ecosystems and occurs as the cupric ion (Cu^{+2}). The effects of copper on aquatic organisms can be directly or indirectly lethal; for example: fish gills become ragged and lose their ability to regulate transport of salts such as sodium chloride and potassium chloride into and out of fish; and Cu adversely affects olfaction (sense of smell) in fish (Solomon 2009). Cu is bioaccumulative in some organisms such as plankton, oyster and squid by factors of up to 10^7 but it does not magnify in the food chain (Rieuwerts 2015).

1.3.3 Iron (Fe)

Fe is the second most abundant metal and fourth most abundant element on the earth's crust (USEPA 1993; Jaishankar et al. 2014) and found in form of iron oxides such as mineral hematite, magnetite, and taconite on the earth's crust. Weathering of rocks and soil, and atmospheric deposition contributes to Fe in the environment naturally. Fe reaches the oceans mainly from rivers as suspended sediment, aeolian dust transport, through hydrothermal activity and by recycling from shelf sediments (Raiswell and Canfield 2012). The transport of aeolian dust over a timeframe of a few days to weeks over large areas of the ocean surface become an efficient means of distributing iron to iron-deficient regions (Emerson 2019). Anthropogenic emissions of Fe include combustion of coal, petroleum, biofuel, fossil fuel and biomass (Guieu et al. 2005; Takahashi et al. 2013; Wang et al. 2015).

In the photic zone, the sources of Fe mostly include wet and dry deposition of atmospheric aerosols, vertical mixing and upwelling and biogenic recycling of cellular iron in surface waters (Wells et al. 1995). Fe is an essential nutrient for the growth of marine phytoplankton in surface waters (Cassar et al. 2007; Raiswell and Canfield 2012; Mead et al. 2013) and thus contributes to ocean primary productivity. Fe modulates marine ecosystems, global carbon cycle and atmospheric carbon dioxide uptake (Luo et al. 2008; Mahowald et al. 2009; Tagliabue et al. 2017). In the seawater, Fe occurs in Fe^{+3} and Fe^{+2} oxidation states. Mahowald et al. (2010) reported that changes in the soluble Fe input to the oceans could have an important impact on oceanic carbon uptake and storage and hence indirectly affect climate.

Fe, along with P and other potential trace elements, is a (co) limiting nutrient of nitrogen (N) fixation in the oligotrophic areas of the ocean (Falkowski 1997; Mills et al. 2004) since nitrogen-fixing organisms (diazotrophs) tend to have higher Fe requirements. The stimulation of these elements during phytoplankton growth and primary production links to carbon sequestration and, consequently, the changing climate throughout the geologic time (Martin 1990; Jickells et al. 2005). Moreover, Bishop et al. (2002) and Boyd et al. (2007) reported that the amount of Fe in remote oceans could increase the production of dimethyl sulfide (DMS) and/or organic carbon from microorganisms in the ocean, which in turn affects the radiative forcing in the atmosphere.

The mineral composition of dust is a key factor in the chemical forms of Fe, and it determines the solubility and thus the bioavailability of Fe (Wang et al. 2015). Cassar et al. (2007) summarized five sources of bioavailable Fe to the Southern Ocean surface waters, i.e., melting of sea ice, the release of dissolved iron or resuspension of sediments, upwelling supplies iron, vertical mixing supplies iron, and delivery of soluble iron by aerosol deposition supplies.

1.3.4 Lead (Pb)

Pb is naturally occurring in the environment (Garcia-Alix et al. 2013; Chiarelli and Roccheri 2014) and a highly toxic metal whose widespread use has caused extensive environmental contamination (Jaishankar et al. 2014). Pb is an extensively used chemical for the preparation of large industry and household-based products (Wang et al. 2014a). Important sources of environmental contamination of Pb include mining, smelting, manufacturing and recycling activities (WHO 2017).

Pb usage is in car batteries, as a coloring element in ceramic glazes, as a base metal for organ pipes, in the glass of computer as well as television screens, household paints (Jaishankar et al. 2014); and hence is largely lost to the atmosphere and eventually finds its way to the marine environment via atmospheric deposition. Wuana and Okieimen (2011), Tchounwou et al. (2012), and Jaishankar et al. (2014) reported that other uses of Pb include house paint (lead based), bullets, solders, bearings, cable covers, ammunition, plumbing, pigments, and caulking.

Leaded gasoline had been a major source of Pb dispersion in the environment. Pb, in the form of tetraethyl lead, was previously added to gasoline to boost the octane levels and protect exhaust valve seats from excessive wear in vehicles since the 1920s (ATSDR 2017). However, their use was phased out in the United States starting in the early 1970s, and banned for use in gasoline for motor vehicles in January 1996 (ATSDR 2017) as a result of actions taken under the Clean Air Act of 1970. The Clean Air Act is the law that defines EPA's responsibilities for protecting and improving the nation's air quality and the stratospheric ozone layer (EPA 2017). The Clean Air Act of 1970 was prompted due to the presence of dense, visible smog in many of the nation's cities and industrial centers. Major revisions were made in 1977 and 1990 to the Act to improve its effectiveness and to target newly recognized air pollution problems such as acid rain and damage to the stratospheric ozone layer (EPA 2017).

Pb is a non-essential element (Jaishankar et al. 2014) and is potentially hazardous to most forms of life. Pb can occur in four different forms in water and these include: highly mobile and bioavailable form (ionic); in a bound form with limited mobility and bioavailability (organic complexes); strongly bound with limited mobility (attached colloidal particles such as iron oxide); and very limited mobility and availability (attached to solid particles of clay or dead remains of organisms) (ILA 2018). Pb does not show biomagnifying characteristics (De Pooter 2013).

1.3.5 Zinc (Zn)

Zn is an essential element with biological significance (Depledge and Rainbow 1990) and found in earth's crust (McComb et al. 2014). It occurs only in the form of sulfide, carbonate and silicate ores, mainly as calamine ($Zn_2SiO_4 H_2O$), smithsonite ($ZnCO_3$), sphalerite (ZnS) and zincite (ZnO) (Namieśnik and Rabajczyk 2010).

Zn compounds are used in the pharmaceuticals industry as ingredients in some common products, such as vitamin supplements, sun blocks, diaper rash ointments, deodorants and antidandruff shampoos (ASTDR 2011; McComb et al. 2014). Zn and Cu are used in shipping activities such as sacrificial anodes and antifoulants (Negri et al. 2002).

Further uses of Zn include galvanizing iron, preparation of certain alloys, roofing and gutter in building construction; whilst zinc oxide is used as an activator in rubber industry, and as a pigment in plastics, cosmetics, photocopier and paper wallpaper and Zn based fungicides (Lenntech 2006).

At low concentrations, Zn contributes to the normal metabolism of most living organisms through its role as a cofactor in > 300 enzymatic reactions (Morel et al. 1994). Zn does not show biomagnification characteristics (De Pooter 2013), excessive levels of Zn in the environment lead to bioaccumulation of Zn in organisms.

1.3.6   Mercury (Hg)

Mercury (Hg) is a toxic, persistent, non-essential metal that exists in the environment in various chemical forms (i.e. elemental, organic and inorganic). Hg is released into the environment from emission sources that can be natural or anthropogenic. These forms of Hg are interconvertible and produce toxicity and bioavailability (Jaishankar et al. 2014). Organic forms of Hg include methylated Hg compounds such as monomethyl mercury (MMHg), dimethyl mercury (DMHg); and phenyl mercury (PhHg). Whereas elemental mercury (Hg^0) is a highly volatile liquid at room temperature and released as a vapour in the atmosphere while inorganic Hg occurs in the form of mercuric ($HgCl_2$, HgS) and mercurous (Hg_2Cl_2) salts. Gaseous elemental Hg has lifetime of about 1 year and once emitted into the atmosphere, it has the potential to be transported on hemispheric and global scales (Slemr et al. 2018).

Hg is persistent in the environment and dispersed over great distances through atmospheric, oceanic, aquatic and biotic transport vectors (Brosset 1981; Bank 2012; Reichelt-Brushett et al. 2017). It is well documented that Hg can easily be transported in the atmosphere as gaseous elemental Hg and deposits to aquatic and terrestrial ecosystems as oxidised and particulate-bound Hg via dry and wet deposition (Selin 2009). The ultimate fate of all Hg emitted to the atmosphere is to deposit to ecosystems (Lyman et al. 2020). Hence, the constant transfer of Hg between the aquatic, atmospheric, biological, and terrestrial reservoirs has made it ubiquitous throughout the environment.

When deposited in anoxic environments, Hg (HgS, HgS_2) is transformed to methyl mercury by microbial systems in the water column, surface sediments and wetlands (Clark 2001; Solomon 2008; Kwon et al. 2020). Organic forms of Hg are more toxic than the inorganic salts (Clark 2001). Methyl Hg (MHg) is the most common form of organic mercury in the environment (Gworek et al. 2016), is a neurodevelopmental toxicant (Obi et al. 2015); and is the most toxic form of Hg (Henriques et al.2015). Clark (2001) reported that an outbreak of MHg poisoning occurred in the small

Japanese coastal town (Minamata disease) due to consumption of contaminated fish and seafood (discussed later in this section).

Natural Hg emission processes include weathering of mercury-bearing rocks, degassing from Hg deposits; degassing from aquatic and terrestrial systems (through reduction of Hg^{2+} to Hg^0); geological activity-volcanic and geothermal processes (underwater exhalations from geothermal vents); biomass burning (forest fires); erosion of mercury –containing minerals and plant growth (Clark 2001; Baby et al. 2010; Okoro et al. 2012; Gworek et al. 2016; Rodrigues et al. 2019).

Anthropogenic uses of Hg include dentistry and in the pharmaceutical industry (Okoro et al. 2012); whilst elemental Hg is used in the extraction of gold (Au) in Artisanal small-scale gold mining (Reichelt-Brushett et al. 2017) and electronics, light bulbs and thermometers. Anthropogenic emissions of Hg result from agricultural practices, municipal wastewater discharge, mining processes, combustion of fossil fuels, electricity-generating power stations, use of Hg in precious metal extraction, manufacture of many commercial products such as thermometers, thermostats, barometers, batteries and dental amalgams and discharges of industrial wastewater (Chen et al. 2012; Driscoll et al. 2013; Jaishankar et al. 2014). When vaporized into the air by factories producing these products, Hg is deposited into soils and later discharged through rivers to the oceans. Anthropogenic emissions of Hg also result from incineration of medical and municipal wastes (Suvarapu et al. 2013). Coal-fired power plants emit Hg as gaseous elemental Hg, gaseous oxidised Hg and particulate bound mercury into the atmosphere (Weigelt et al. 2016).

The health hazards of Hg became well-known and gained worldwide attention due to the 1956 outbreak of Minamata disease in Japan (Ye et al. 2016). Methylmercury, formed as a by-product of chemical compound (acetaldehyde) production in a fertilizer factory was discharged into the Minamata Bay for a long time; thus polluting the marine ecosystem including the fish and shellfish. Upon consumption of the contaminated seafood, many residents from the area died or suffered injury to their nervous system, brain, heart, lungs, kidneys and the immune system. Due to its adverse health problems, the World Health Organization (WHO) considers Hg as one of the top 10 chemicals of major public health concern.

Methyl mercury has biomagnifying characteristics, while inorganic Hg does not. Methylmercury dissolves well in water, crosses biological membranes and persists in fatty tissues of organisms (Solomon 2008). Warmer water temperatures increase bacterial Hg methylation and uptake by plankton and zooplankton (Costa et al. 2012).

1.3.7 Nickel (Ni)

Ni is a naturally occurring element that can exist in various mineral forms and is in global demand as an industrially and commercially important commodity (Cempel and Nikel 2006) due to its versatility and use as an alloying ingredient. Ni is used as an alloying component in stainless steel, cast iron, nonferrous alloys and other nickel-based alloys used in challenging high temperature and corrosion resistant applications (Reichelt-Brushett and Hudspith 2016). Further prominent uses of Ni include Ni coatings, components in battery systems and electrodes, coinage, filters and as catalysts in commercially significant chemical processes (see Reichelt-Brushett and Hudspith 2016).

Ni ore occurs as either magmatic sulphides or nickel laterites (Mudd 2010; Gissi et al. 2017; Gillmore et al. 2020) with Ni laterite ore being extracted by opencast mining, thus leading to major landscape alteration, enhanced soil erosion rates, and the production of large volumes of waste material (Gillmore et al. 2020), which later enters the marine environment.

Ni is released from both natural and anthropogenic sources; with input from both stationary and mobile sources (Cempel and Nikel 2006). These sources include weathering of rocks and soils, volcanic emissions, forest fires and vegetation, fossil fuels (usually rich in Ni), combustion of oil and coal and incineration of waste and sewage (Clark 2001; Cempel and Nikel 2006). Ni is a substantial contaminant in the marine sediments of industrialized areas and has been used in steel, batteries and as a catalyst. The major input of Ni to the sea is through rivers (Clark 2001). Ni does not show any biomagnification characteristics and is regarded as a moderately toxic metal. It can cause lethal effects to some algae species at concentrations above 600 μg/l (Clark 2001).

1.3.8 Chromium (Cr)

Cr is a naturally occurring element present in the earth's crust and enters the environment from a wide variety of natural and anthropogenic sources (Tchounwou et al. 2012). Wuana and Okieimen (2011) mentioned that Cr is mined as a primary ore product in the form of chromite, $FeCr_2O_4$. Sources for Cr in the oceans are mineral weathering processes and riverine and atmospheric input (Geisler and Schmidt 1991; Bielicka et al. 2005).

Industries with the largest contribution to Cr release in the environment include metal processing, tannery facilities, chromate production, stainless steel welding, and ferrochrome and chrome pigment production (Tchounwou et al. 2012) and textile dyes, catalysts, wood and water treatment (Bielicka et al. 2005). Cr is a highly toxic metal presenting various degrees of risk for coastal ecosystems mainly due to dumping of untreated or poorly treated industrial residues (Chiarelli and Roccheri 2014).

Cr exists in variable oxidation states but only $+3$ and $+6$ oxidation states are biologically and environmentally stable (Ducros 1992) and the hexavalent form of Cr is 30 times more toxic than the trivalent form and can be mutagenic and carcinogenic (Natale et al. 2000; Chiarelli and Roccheri 2014).

1.3.9   Tin (Sn)

Sn occurs naturally in the earth's crust and is found in various environmental media in both inorganic and organic forms. Sn metal is used to line cans for food, beverages, and aerosols (ATSDR 2005). Inorganic Sn compounds (e.g. stannous chloride, stannous sulphide, and stannic oxide) are used in toothpaste, perfumes, soaps, food additives and dyes; whilst organic tin compounds (e.g. dibutyltin, tributyltin, triphenyltin) are used to make plastics, food packages, plastic pipes, pesticides, paints and insect repellents (ATSDR 2005).

Sn enters the environment by both natural processes as well as through anthropogenic activities such as mining, coal and oil combustion, and the production and use of tin compounds (ATSDR 2005). Soil dust and forest fires are the natural sources of atmospheric Sn emissions (Byrd and Andreae 1982). Sn enters the aquatic environment through atmospheric deposition, riverine input, and sediment resuspension (Duan et al. 2012). Once in the environment, metallic Sn quickly converts to inorganic tin compounds whilst inorganic Sn only changes its forms and organic tin compounds are degraded into inorganic Sn compounds by sunlight and bacteria (ATSDR 2005).

Organotin compounds have been extensively used in boat paint since 1960 because of their excellent and long lasting antifouling properties (Okoro et al. 2011). Tributyltin (TBT) and tributyltin fluoride (TBTF) are used in antifouling paints as a finish coat to the immersed sections of boats and floating structures. Hence, there is a possibility of global dispersion of TBT throughout the marine environment, from the coastal zone to the open ocean (de Mora 1999).

TBT is extremely surface active and therefore readily adsorbed onto suspended particulate material (de Mora 1999). TBT degrades; either through photochemical reactions or microbial mediated pathways; in a process referred to as debutylation in seawater to give nontoxic compounds dibutyltin and monobutyltin (de Mora 1999; Maata and Koshy 2001). Despite being the most widely used active component in antifouling paints, a global ban was ratified in 2008 on the application of TBT paints (Tornero and Hanke 2016).

TBT is extremely toxic and lethal to a variety of planktonic organisms (Clark 2001); and thus results in a wide range of deleterious biological effects on non-target organisms (de Mora 1999). It has an endocrine disrupting effect, particularly on shellfish (Dafforn et al. 2011). de Mora (1999) reported that concentrations as low as $0.02 \, \mu g$ TBT-Sn L^{-1} is lethal to some shellfish; and lower concentrations result in sub-lethal effects such as poor growth rates and reduced recruitment leading to the decline of shellfisheries.

Sn is an essential element for organisms yet is a toxic cumulative element and moderate level of Sn may impede the growth of organism and lead to harmful environmental effects (Duan et al. 2012). Sn toxicity grows with the augmenting of both pH values and duration in aquatic systems (Duan et al. 2012), thus the bioavailability of Sn is highest at neutral and slightly alkaline pH and reduced in the presence of humic acid (Pawlik-Skowrońska et al. 1997).

1.3.10 Arsenic (As)

As is a naturally occurring metal found in the earth's crust, occurring either as elemental (metallic), inorganic, or as organic As (ATSDR 2007). As is a toxic metalloid that exists in two oxidative states: a trivalent form (in the form of arsenous acid and its salts) and a pentavalent form (in the form of arsenic acid and its salts) (Chiarelli and Roccheri 2014). The toxicity of As depends upon the nature of the compound formed and its valency, with the trivalent form being more toxic than the pentavalent form (Clark 2001).

Kaur et al. (2011) reported that As is one of the most important global environmental pollutant and a persistent bioaccumulative carcinogen. Environmental pollution by As occurs as a result of natural phenomena such as volcanic eruptions and soil erosion, and through anthropogenic activities (ATSDR 2000). The natural inputs of As to the sea is chiefly by rivers especially from areas of metalliferrous mining (Clark 2001; Nair et al. 2003). Anthropogenic sources of As include the smelting of Cu, Ni, Pb and Zn ores, burning of fossil fuels in households and power plants, growth stimulants for plants and animals (Chiarelli and Roccheri 2014).

Furthermore, Tchounwou et al. (2012) reported that there are several As-containing compounds produced industrially, which have been used to manufacture products with agricultural applications such as insecticides, herbicides, fungicides, algicides, sheep dips, wood preservatives, and dye-stuffs. El-Sorogy et al. (2016) also reported that the largest source of pollution from As result from agricultural chemicals such as herbicides, fungicides, rodenticides and insecticides.

Marine organisms contain As especially in the form of arsenobutaine (which is pentavalent) and thus very stable, metabolically inert and non-toxic; whilst marine algae contain arsenic as carbohydrate compounds (Clark 2001). Even though a variety of aquatic organisms, including some algae, crustaceans, and fish, bioaccumulate arsenic, it does not magnify through the food chain (Clark 2001).

1.3.11 Silver (Ag)

Ag is a rare but naturally occurring metal, often found deposited as a mineral ore in association with other elements (Howe and Dobson 2002). Argentite, nickel ores, lead-zinc and porphyry copper ores, and platinum and gold deposits are used to extract Ag, however, other sources of Ag include new scrap generated in the manufacture of silver-containing products; coin and bullion; and old scrap from electrical products, old film and photoprocessing wastes, batteries, jewellery, silverware, and bearings (Howe and Dobson 2002).

Anthropogenic sources of Ag in the environment include emissions from smelting operations; manufacture and disposal of certain photographic and electrical supplies, coal combustion, industrial and municipal wastewater outfalls and cloud seeding

(Eisler 1997; Purcell and Peters 1998) and these have adversely impacted the environment.

Ag is extremely toxic in its ionic form (Ag^+) (Luoma et al. 1995) to marine phytoplankton and invertebrates, and its increasing use as a biocide has raised concerns about its potential as an environmental pollutant (Purcell and Peters 1998). Luoma (2008) and Fabrega et al. (2011) have reported that Ag^+ ions have a great propensity to bioconcentrate in organisms since the chemical properties of Ag^+ ions make them compatible for uptake via cell membrane ion transporters. Ag can be bioaccumulated from solution by phytoplankton, some algae and oysters, fish, shrimp and a variety of gastropods (Clark 2001). Ag in sediments is bioavailable and despite being bioaccumulated by a wide variety of animals, biomagnification is not evident (Clark 2001).

1.3.12   Manganese (Mn)

Mn is one of the most abundant and broadly distributed metals in nature (Pinsino et al. 2012); occurring in the Earth's crust in ores such as pyrolusite (MnO_2), rhodocrosite ($MnCO_3$), manganite ($Mn_2O_3.H_2O$), hausmannite (Mn_3O_4), biotite mica (K (Mg, Fe)$_3$ ($AlSi_3O_{10}$) $(OH)_2$ (Moore 1991). Mn is the 12th most abundant element in the earth's crust and the fifth most plentiful metal (Röllin 2011).

Hansel (2017) reported Mn is widely distributed throughout the global ocean as an essential antioxidant (Mn^{2+}), a potent oxidant (Mn^{3+}) and strong adsorbent (Mn oxides) sequestering disproportionately high levels of trace metals and nutrients in comparison to the surrounding seawater. Dissolved Mn in oceanic waters range from 0.2 to 5.0 nmol/kg (Graham et al. 1988). Mn exhibits unique redox dynamics as Mn^{+2}, Mn^{+3}, Mn^{+4}; with Mn^{+4} being the most abundant form found in minerals (Burdige 1993, Fischel et al. 2015). In fact, Mn is one of the key considerations used as a tracer of ocean processes.

Mn is regarded as a biologically essential trace metal, a major global commodity and an emerging marine contaminant for which ecotoxicological data are inadequate (ANZECC and ARMCANZ 2016); with the oxidation of soluble Mn^{2+} to insoluble Mn^{4+} form having significant ecotoxicological consequences (Summer et al. 2019). Mn^{2+}, being bioavailable and potentially toxic to marine organisms for prolonged periods of time has high solubility and slow transformation rates to insoluble forms (Pinsino et al. 2012; Summer et al. 2019).

Ocean spray, forest fires, vegetation, crustal rock and volcanic activity are the major natural atmospheric sources of Mn (Röllin 2011). Anthropogenic sources of Mn include mining, industrial emissions, fossil fuel combustion, and erosion of manganese-containing soils. Hagelstein (2009) mentioned that Mn metal is vital to metallurgical industries such as fabrication of steel and Al; with electrolytic Mn being a constituent of nonferrous metals improving their strength and ductility. In addition, Mn usage is also in the production of dry-cell batteries, plant fertilizer components, animal feed and colorant for bricks (Hagelstein 2009).

1.4 Metal Speciation

Metal speciation refers to the various physical and chemical forms in which an element may exist in the marine environment; and hence this determines the behaviour of metals (Ansari et al. 2004). Speciation influences metal bioavailability and toxicity to biota, its transportation and mobilization, and its interaction with the environment. An understanding of chemical speciation and bioavailability becomes paramount in the study of fate and effects of trace metals in the environment (VanBriesen et al. 2010). Upon dissolution and reaction with anions available in seawater, heavy metals give a number of chemical species at different pH's (Ansari et al. 2004), thus influencing metal bioavailability and toxicity. Factors such as pH, the types and concentrations of inorganic and organic ligands, and the colloidal species present influence the actual metal speciation (Violante et al. 2010). The formation of metal-organic complexes influences the speciation of many heavy metal ions, which have the tendency to bind with natural organic compounds such as fulvic acids or with colloidal matter.

Elder (1988) reported that the toxicity of a given trace metal is intimately linked to chemical speciation, which is governed by the physico-chemical nature of seawater and can be variable in dynamic coastal waters. Metal loads tend to be higher in coastal marine waters due to surrounding landmasses and associated human activities (Furness and Rainbow, 1990).

Libes (1992) mentioned that one of the keys to considering metal toxicity is understanding the effects of the physico-chemical characteristics of water on metal speciation and bioavailability, as not all forms of metals are equally toxic. Heavy metal toxicity is dependent upon chemical speciation (Ansari et al. 2004). The toxicity data obtained contribute to the development of risk assessment tools such as water quality guidelines and bioavailability-based models (Gissi et al. 2017). Governments and relevant stakeholders use the water quality guidelines to set thresholds indicating potential risks to aquatic environments from exposure to contaminants such as metals (Wang et al. 2014b) and is in fact a good initiative as specific water quality guidelines would be set for all heavy metal pollutants.

Some analytical techniques used in heavy metal speciation in fresh, estuarine and marine environments include anodic stripping voltammetry, ultrafiltration, ion exchange or chelating resins, dialysis and UV irradiation.

1.5 Summary

Being natural constituents of the earth's crust, heavy metals generally are in very low concentrations; however, human activities have unavoidably increased the metal concentrations in the marine environment. When present in low concentrations, heavy metals are quintessential in maintaining various biochemical and physiological functions in living organisms, however, when certain threshold concentrations exceed, they become noxious.

Heavy metals have various routes of entry into the marine environment including industrial wastewater discharge; storm water runoff, agricultural activities, dredging activities, atmospheric deposition and sewage discharge.

Heavy metals have been observed as continuous additions to the marine environment; non-biodegradable; persistent and toxic and do not get broken down by bacterial action. In presence of acidic mediums, the toxicity of the heavy metals increases. This has led to serious ecotoxicological problems for marine species.

Metal speciation influences the bioavailability and toxicity of the heavy metals to the aquatic biota. Metal ions can form complexes with organic and inorganic ligands as well as colloidal species and enter living biota via processes like sorption or biological uptake.

References

Agency for Toxic Substances and Disease Registry (ATSDR) (2000) Toxicological profile for Arsenic TP-92/09. Georgia: Center for Disease Control, Atlanta. https://www.atsdr.cdc.gov/tox profiles/tp2.pdf

Agency for Toxic Substances and Disease Registry (ATSDR) (2005) Toxicological profile for tin and compounds. https://www.atsdr.cdc.gov/toxprofiles/tp55.pdf

Agency for Toxic Substances and Disease Registry (ATSDR). (2007) Toxicological profile for Arsenic. Atlanta, GA: U.S. Department of Health and Human Services, Public Health Service. https://www.atsdr.cdc.gov/toxprofiles/tp2.pdf

Agency for Toxic Substances and Disease Registry (ATSDR) (2011) Toxicology profile for arsenic. TP-92/09. Center for Disease Control, Agency for Toxic Substances and Disease Registry; Atlanta, GA

Agency for toxic substances and disease registry (ATSDR) (2017) Lead toxicity. https://www.atsdr. cdc.gov

Akan JC, Abdurrahman FI, Sodipo OA, Ochanya AE, Askira YK (2010) Heavy metals in sediments from River Ngada, Maiduguri Metropolis, Borno State, Nigeria. J Environ Chem Ecotoxicol 2:131–140

Alexander DE (1999) Bioaccumulation, bioconcentration, biomagnification. Environ Geol Encycl Earth Sci: 43–44. https://doi.org/10.1007/1-4020-4494-1_31

Al-Rousan S (2012) Skeletal extension rate of the reef building coral *Porites* species from Aqaba and their environmental variables. Nat Sci 4(9):731–739

Al-Rousan SA, Al-Shoul RN, Al-Horani FA, Abu-Hilal AH (2007) Heavy metal contents in growth bands of *Porites* corals. Record of anthropogenic and human developments from the Jordanian Gulf of Aqaba. Mar Pollut Bull 54:1912–1922

ANZECC & ARMCANZ (Australia and New Zealand Environment and Conservation Council & Agriculture and Resource Management Council of Australia and New Zealand (2016) Default guideline values for toxicants: manganese-marine water (DRAFT). Revision to the Australian and New Zealand guidelines for fresh and marine water quality. Australian Department of Agriculture and Water Resources, Canberra. Unpublished, p 17

Ansari TM, Marr IL, Tariq N (2004) Heavy metals in marine pollution perspective—a mini review. J Appl Sci 4(1):1–20

Ashraf W (2005) Accumulation of heavy metals in kidney and heart tissues of *Epinephelus Micrdon* fish from the Arabian Gulf. Environ Monit Assess 101(1–3):311–316

Ayangbenro AS, Babalola OO (2017) A new strategy for heavy metal polluted environments: a review of microbial biosorbents. Int J Environ Res Public Health 14(94). https://doi.org/10.3390/ijerph14010094

Baby J, Raj JS, Biby ET, Sankarganesh P, Jeevitha MV, Ajisha SU Rajan SS (2010) Toxic effect of heavy metals on aquatic environment. Int J Biolog Chem Sci 4(4):939–952

Bank MS (2012) Mercury in the environment: pattern and process, 1st edn. University Of California Press, pp 340

Bastami KD, Neyestani MR, Shemirani F, Soltani F, Haghparast S, Akbari A (2015) Heavy metal pollution assessment in relation to sediment properties in the coastal sediments of the Southern Caspian Sea. Mar Pollut Bull 93:237–243

Bastidas C, Bone D, Garci'a EM (1999) Sedimentation rates and metal content of sediments in a Venezuelan coral reef. Mar Pollut Bull 38:16–24

Bat L, Sezgin M, Ustun F, Sahin F (2012) Heavy metal concentrations in ten species of fishes caught in Sinop coastal waters of the Black Sea, Turkey. Turkish J Fish Aquatic Sci 12(5):371–376. https://doi.org/10.4194/1303-2712-v12_2_24

Bielicka A, Bojanowska I, Wisniewski A (2005) Two faces of chromium-pollutant and bioelement: review. Polish J Environ Stud 14(1):5–10

Bielmyer GK, Grosell M (2011) Emerging issues in marine metal toxicity. In: Bury N, Handy R (eds) Essential reviews in experimental biology, vol 2. Kings College. London, UK, pp 129–158

Bielmyer GK, Brix KV, Capo TR, Grosell M (2005) The effects of metals on embryo larval and adult life stages of the sea urchin, *Diadema antillarum*. Aquat Toxicol 74:254–263

Bielmyer GK, Gillette P, Grosell M, Bhagooli R, Baker AC, Langdon C, Capo T (2010) Effects of copper exposure on three species of scleractinian corals. Aquat Toxicol 97:125–133

Bielmyer GK, Jarvis T, Harper BT, Butler B, Rice L, Ryan S (2012) Metal accumulation from dietary exposure in the sea urchin, *Strongylocentrotus droebachiensis*. Arch Environ Contam Toxicol 63:86–94

Bielmyer-Fraser GK, Patel P, Capo T, Grosell M (2018) Physiological responses of corals to ocean acidification and copper exposure. Mar Pollut Bull 133:781–790

Bishop JKB, Davis RE, Sherman JT (2002) Robotic observations of dust storm enhancement of carbon biomass in the North Pacific. Science 298:817–821

Boyd P, Jickells TD, Law CS, Blain S, Boyle EA, Buesseler KO, Coale KH, Cullen JJ, de Baar HJW, Follows M, Harvey M, Lancelot C, Levasseu M, Owens NPJ, Pollard R, Rivkin RB, Sarmiento J, Schoemann V, Smetacek V, Takeda S, Tsuda A, Turner S, Watson AJ (2007) Mesoscale iron enrichment experiments 1993-2005: synthesis and future directions. Science 315:612–617

Brosset C (1981) The mercury cycle. Water Air Soil Pollution 16(2):253–255. https://doi.org/10.1007/bf01046859

Brown MT, Depledge MH (1985) Determinants of trace metal concentrations in marine organisms. In: Langston WJ, Bebianno MJ (eds) Metal metabolism in aquatic environment

Brown CJ, Eaton RA (2001) Toxicity of chromated arsenate (CCA) treated wood to non-target marine fouling communities in Langstone Harbour, Portsmouth, UK. Mar Pollut Bull 42(4):310–318

Burdige DJ (1993) The biogeochemistry of manganese and iron reduction in marine sediments. Earth-Sci Rev 35:249–284. https://doi.org/10.1016/0012-8252(93)90040-E

Byrd JT, Andreae MO (1982) Tin and methyltin species in seawater: concentrations and fluxes. Science 218(5):565–569

Cassar N, Bender ML, Barnett BA, Fan S, Moxim WJ, Levy H, Tilbrook B (2007) The southern ocean biological response to aeolian iron deposition. Science 317:1067–1070

Cempel M, Nikel G (2006) Nickel: A Review of its Sources and Environmental Toxicology. Polish J Environ Stud 15(3):375–382

Chen CW, Chen CF, Dong CD (2012) Distribution and accumulation of mercury in sediments of Kaohsiung River Mouth, Taiwan. APCBEE Procedia 1:153–158

Chen X, Wei G, Deng W, Liu Y, Sun Y, Zeng T, Xie L (2015) Decadal variations in trace metal concentrations on a coral reef: evidence from a 159-year record of Mn, Cu, and V in a Porites coral from the northern South China Sea. J Geophys Res Oceans 120:405–416. https://doi.org/10.1002/2014JC010390

Chiarelli R, Roccheri MC (2014) Marine invertebrates as bioindicators of heavy metal pollution. Open J Metal 4:93–106. https://doi.org/10.4236/ojmetal.2014.44011

Clark RB (2001) Marine pollution. Oxford University Press

Claisse D, Alzieu C (1993) Copper contamination as a result of antifouling regulation. Mar Pollut Bull 26:395–397

Costa MF, Landing WM, Kehrig HA, Barletta M, Holmes CD, Barrocas PRG, Evers DC, Vasconcellos AC, Hacon SS, Moreira JC, Malm O (2012) Mercury in tropical and subtropical coastal environments. Environ Res 119:88–100

Cremlyn R (1979) Pesticides preparation and mode of action. Wiley, London

Crespo S, Karnaky KJ Jr (1983) Copper and zinc inhibit chloride transport across the opercular epithelium of seawater-adapted killifish *Fundulus heteroclitus*. J Exp Biol 102:337–341

Dafforn KA, Lewis JA, Johnston EL (2011) Antifouling strategies: history and regulation, ecological impacts and mitigation. Mar Pollut Bull 62:453–465

Das S, Patnaik SC, Sahu HK, Chakraborty A, Sudarshan M, Thatoi HN (2013) Heavy metal contamination, physio-chemical and microbial evaluation of water samples collected from chromite mine environment of Sukinda, India. Trans Nonferr Metals Soc China 23:484–493

DeForest D, Brix K, Adams W (2007) Assessing metal bioaccumulation in aquatic environments: The inverse relationship between bioaccumulation factors, trophic transfer factors and exposure concentration. Aquat Toxicol 83:236–246

Depledge MH, Rainbow PS (1990) Models of regulation and accumulation of trace metals in marine invertebrates. Comp Biochem Physiol Pat C Comp Pharmacol 97(1):1–7

de Mora SJ (1999) The oceanic environment. In: RM Harrison (ed) Understanding our environment: an introduction to environmental chemistry and pollution, p 195

De Pooter D (2013) Heavy metals. http://www.coastalwiki.org/wiki/Heavy_metals

Driscoll CT, Mason RP, Chan HM, Jacob DJ, Pirrone N (2013) Mercury as a global pollutant: sources, pathways and effects. Environ Sci Technol 47:4967–4983. https://doi.org/10.1021/es305071v

Duan L, Song J, Li X, Yuan H, Xu S (2012) Dissolved inorganic tin sources and its coupling with eco-environments in Bohai Bay. Environ Monit Assess 184:1335–1349. https://doi.org/10.1007/s10661-011-2044-4

Ducros V (1992) Chromium metabolism, a literature review. Biol Trace Elem Res 32:65–77

Eisler R (1997) Silver hazards to fish, wildlife and invertebrates: a synoptic review. Washington, DC, US Department of the Interior, National Biological Service, 44 pp. (Biological Report 32 and Contaminant Hazard Reviews Report 32)

Elder JF (1988) Metal biogeochemistry in surface-water systems; a review of principles and concepts. US Geological Survey Circular, pp 1013

El-Sorogy AS, Youssef M, Al-Kahtany K, Al-Otaiby N (2016) Assessment of arsenic in coastal sediments, seawaters and molluscs in the Tarut Island, Arabian Gulf, Saudi Arabia. J Afr Earth Sc 113:65–72

Emerson D (2019) Biogenic Iron Dust: a novel approach to ocean iron fertilization as a means of large scale removal of carbon dioxide from the atmosphere. Front Mar Sci 2:22. https://doi.org/10.3389/fmars.2019.00022

Environmental Protection Agency (2017) https://www.epa.gov/clean-air-act-overview/clean-air-act-text

Esslemont G (1998) Heavy metals in the tissues and skeleton of *scleractinian* corals. PhD thesis. Southern Cross University, Australia, p 257

Fabrega J, Luoma SN, Tyler CR, Galloway TS, Lead JR (2011) Silver nanoparticles: behaviour and effects in the aquatic environment. Environ Int 37(2):517–531. https://doi.org/10.1016/j.envint.2010.10.012

Falkowski PG (1997) Evolution of the nitrogen cycle and its influence on the biological sequestration of CO_2 in the ocean. Nature 387:272–275

Farombi EO, Adelowo OA, Ajimoko YR (2007) Biomakers of oxidative stress and heavy metal levels as indicators of environmental pollution in African cat fish (Clarias gariepinus) from Nigeria Ogun River. Int J Environ Res Public Health 4(2):158–165

Fischel JS, Fischel MH, Sparks DL (2015) Advances in understanding reactivity of manganese oxides with arsenic and chromium in environmental systems. In: ACS symposium series. Oxford University Press, Washington DC, pp 1–27. https://doi.org/10.1021/bk-2015-1197.ch001

Fu F, Wang Q (2011) Removal of heavy metal ions from wastewaters: a review. J Environ Manage 92:407–418

Furness RW, Rainbow PS (1990) Heavy metals in the marine environment. CRC Press, Boca Raton, FL

Garcia-Alix A, Jimenez-Espejo FJ, Lozano JA, Jiménez-Moreno G, Martinez-Ruiz F, Garcia Sanjuán L, Aranda Jiménez G, García Alfonso E, Ruiz-Puertas G, Scott Anderson R (2013) Anthropogenic impact and lead pollution throughout the Holocene in Southern Iberia. Sci Total Environ 449:451–460

Gautam RK, Sharma SK, Mahiya S, Chattopadhyaya MC (2014) Contamination of heavy metals in aquatic media: transport, toxicity and technologies for remediation (Chapter 1). In: Sharma SK (ed) Heavy metals in water: presence. Removal and safety

Geisler CD, Schmidt D (1991) An overview of chromium in the marine environment. Ocean Dyn 44(4):185–196. https://doi.org/10.1007/BF02226462

Giga W (2009) The Rhine red, the fish dead-the 1986 Schweizerhalle disaster, a retrospect and long-term impact assessment. Environ Sci Pollut Res Int

Gillmore ML, Gissi F, Golding LA, Stauber JL, Reichelt-Brushett AJ, Severati A, Humphrey CA, Jolley DF (2020) Effects of dissolved nickel and nickel-contaminated suspended sediment on the scleractinian coral, Acropora muricata. Mar Pollut Bull 152: https://doi.org/10.1016/j.marpolbul.2020.110886

Gissi F, Stauber J, Reichelt-Brushett AJ, Harrison PL, Jolley DF (2017) Inhibition in fertilisation of coral gametes following exposure to nickel and copper. Ecotoxicol Environ Saf 145:32–41

Graham RD, Hannam RJ, Uren NC (1988) Manganese in soils and plants, manganese in soils and plants. Springer Netherlands, Dordrecht. https://doi.org/10.1007/978-94-009-2817-6

Grosell M (2011) Copper. In: Wood, Farrell, Brauner (eds) Fish physiology. Homeostasis and toxicity of essential metals, vol 31, pp 54–110

Guieu C, Bonnet S, Wagener T, Loÿe-Pilot M-D (2005) Biomass burning as a source of dissolved iron to the open ocean? Geophys Res Lett 32:L19608

Guzma'n HM, Jime'nez CE (1992) Contamination of coral reefs by heavy metals along the Caribbean Coast of Central America (Costa Rica and Panama). Mar Pollut Bull 24:554–561

Gworek B, Bemowska-Kalabun O, Kijeńska M, Wrzosek-Jakubowska J (2016) Mercury in marine and ocean waters-a review. Water Air Soil Pollution 227:371. https://doi.org/10.1007/s11270-016-3060-3

Hagelstein K (2009) Globally sustainable manganese metal production and use. J Environ Manage 90:3736–3740. https://doi.org/10.1016/j.jenvman.2008.05.025

Hansel CM (2017) Manganese in Marine Microbiology. Adv Microb Physiol 70:37–83. https://doi.org/10.1016/bs.ampbs.2017.01.005 Epub 2017 Mar 14 PMID: 28528651

Harada M (1995) Minamata disease: methylmercury poisoning in Japan caused by environmental pollution. Crit Rev Toxicol 25(1):1–24

Haynes D, Christie C, Marshall P, Dobbs K (2002) Antifoulant concentrations at the site of the Bunga Teratai Satu grounding, Great Barrier Reef, Australia. Mar Pollut Bull 44:968–972

Henriques B, Rocha LS, Lopes CB, Figueira P, Monteiro RJR, Duarte AC, Pardal MA, Pereira E (2015) Study on bioaccumulation and biosorption of mercury by living marine macroalgae: prospecting for a new remediation biotechnology applied to saline waters. Chem Eng J 281:759–770

Hooda PS (2010) Trace elements in soils. Wiley Publication

Hosono T, Su C, Delinom R, Umezawa Y, Toyota T, Kaneko S, Taniguchi M (2011) Decline in heavy metal contamination in marine sediments in Jakarta Bay, Indonesia due to increasing environmental regulations. Estuar Coast Shelf Sci 92:297–306

Howe PD, Dobson S (2002) Silver and silver compounds: Environmental aspects. Concise International Chemical Assessment Document 44. WHO Geneva. https://apps.who.int/iris/bitstream/10665/42553/1/9241530448.pdf

Howe PL, Reichelt-Brushett AJ, Clark MW (2014) Investigating lethal and sublethal effects of the trace metals cadmium, cobalt, lead, nickel and zinc on the anemone *Aiptasia pulchella*, a cnidarian representative for ecotoxicology in tropical marine environments. Mar Freshw Res 65:551–561. CSIRO Publishing. http://dx.doi.org/10.1071/MF13195

Hudspith M, Reichelt-Brushett AJ, Harrison PL (2017) Factors affecting the toxicity of trace metals to fertilization success in broadcast spawning marine invertebrates: a review. Aquat Toxicol 184:1–13. https://doi.org/10.1016/j.aquatox.2016.12.019

International Lead Association (2018) Lead in aquatic environments: understanding the science. https://www.ila-lead.org/responsibility/lead-in-aquatic-environments–understanding-the-science

Jaishankar M, Tseten T, Anbalagan N, Mathew BB, Beeregowda KN (2014) Toxicity, mechanism and health effects of some heavy metals. Interdiscip Toxicol 7(2):60–72. https://doi.org/10.2478/intox-2014-0009

Jickells TD, An ZS, Andersen KK, Baker AR, Bergametti G, Brooks N, Cao JJ, Boyd PW, Duce RA, Hunter KA, Kawahata H, Kubilay N, La Roche J, Liss PS, Mahowald N, Prospero JM, Ridgwell AJ, Tegen I, Torres R (2005) Global iron connections between desert dust, ocean biogeochemistry, and climate. Science 308:67–71

Jones RJ (2007) Chemical contamination of a coral reef by the grounding of a cruise ship in Bermuda. Mar Pollut Bull 54:905–911

Kabata-Pendia A (2001) Trace elements in soils and plants, 3rd edn. CRC Press, Boca Raton, FL

Kaur S, Kamli MR, Ali A (2011) Role of arsenic and its resistance in nature. Can J Microbiol 57:769–774. https://doi.org/10.1139/w11-062

Kim BE, Nevitt T, Thiele DJ (2008) Mechanisms for copper acquisition, distribution and regulation. Nat Chem Biol 4:176–185

Kim BM, Rhee JS, Jeong CB, Seo JS, Park GS, Lee YM, Lee JS (2014) Heavy metals induce oxidative stress and trigger oxidative stress-mediated heat shock protein (*hsp*) modulation in the intertidal copepod *Tigriopusjaponicus*. Comp Biochem Physiol C Toxicol Pharmacol 66:65–74. https://doi.org/10.1016/j.cbpc.2014.07.005

Klaassen CD, Liu J, Diwan BA (2009) Metallothionein protection of Cadmium toxicity. Toxicol Appl Pharmacol 283(3):215–220. https://doi.org/10.1016/j.taap.2009.03.026

Kroon FJ, Berry KLE, Brinkman DL, Kookana R, Leusch FDL, Melvin SD, Neale PA, Negri AP, Puotinen M, Tsang JJ, van de Merwe JP, Williams M (2020) Sources, presence and potential effects of contaminants of emerging concern in the marine environments of the Great Barrier Reef and Torres Strait, Australia. Sci Total Environ 719 (135140). https://doi.org/10.1016/j.scitotenv.2019.135140

Kwon SY, Blum JD, Yin R, Tsui MTK, Yang YH, Choi JW (2020) Mercury stable isotopes for monitoring the effectiveness of the Minamata Convention on Mercury. Earth Sci Rev 203: https://doi.org/10.1016/j.earscirev.2020.103111

Lata R and Rohindra D (2002) Heavy Metals. In: Khurma J, Sutcliffe S (eds) Major environmental issues: an outreach to the South Pacific

Lenntech (2006) Heavy metals. http://www.lenntech.com/heavy-metals.htm

Libes S (1992) An introduction to marine biogeochemistry. Wiley, Singapore

Loganathan P, Hedley MJ (1997) Downward movement of cadmium and phosphorous from phosphate fertilizers in a pasture soil in New Zealand. Environ Pollut 95:319–324. https://doi.org/10.1016/s0269-7491(96)00142-x

Luo C, Mahowald N, Bond T, Chuang PY, Artaxo P, Siefert R, Chen Y, Schauer J (2008) Combustion iron distribution and deposition. Glob Biogeochem Cycles 22, GB1012. https://doi.org/10.1029/2007GB002964

Luoma SN (2008) Silver nanotechnologies and the environment: old problems and new challenges?. Woodrow Wilson International Centre for Scholars or the PEW Charitable Trusts, Washington DC

Luoma SN, Ho YB, Bryan GW (1995) Fate, bioavailability and toxicity of silver in estuarine environments. Mar Pollut Bull 31:44–54

Luoma SN, Rainbow PS (2005) Why is metal bioaccumulation so variable? Biodynamics as a unifying concept. Environ Sci Technol 39(7):1921–1931

Lyman SN, Cheng I, Gratz LE, Weiss-Penzias P, Zhang L (2020) An updated review of atmospheric mercury. Sci Total Environ 707(135575). https://doi.org/10.1016/j.scitotenv.2019.135575

Maata M, Koshy K (2001) A study on tributyltin contamination of marine sediments in the major ports of Fiji. South Pac J Nat Sci 19(1):1–4

Mahowald NM, Engelstaedter S, Luo C, Sealy A, Artaxo P, Benitez-Nelson C, Bonnet S, Chen Y, Chuang PY, Cohen DD, Dulac F, Herut B, Johansen AM, Kubilay N, Losno R, Maenhaut W, Paytan A, Prospero JM, Shank LM, Siefert RL (2009) Atmospheric iron deposition: Global distribution, variability and human perturbations. Ann Rev Mar Sci 1(1):245–278

Mahowald NM, Kloster S, Engelstaedter S, Moore JK, Mukhopadhyay S, McConnell JR, Albani S, Doney SC, Bhattacharya A, Curran MAJ, Flanner MG, Hoffman FM, Lawrence DM, Lindsay K, Mayewski PA, Neff J, Rothenberg D, Thomas E, Thornton PE, Zender CS (2010) Observed 20th century desert dust variability: impact on climate and biogeochemistry. Atmos Chem Phys 10:10875–10893

Main WPL, Ross C, Bielmyer GK (2010) Copper accumulation and oxidative stress in the sea anemone, *Aiptasia pallida*, after waterborne copper exposure. Comp Biochem Physiol Part C 151:216–221

Manahan SE (2003) Toxicological chemistry and biochemistry. Lewis Publishers, CRC Press Company, Boca Raton-London-New York-Washington

Mance G (1987) Pollution threat of heavy metals in aquatic environments. Elsevier, Amsterdam, p 372

Mani D, Kumar C (2014) Biotechnological advances in bioremediation of heavy metals contaminated ecosystems: An overview with special reference to phytoremediation. Int J Environ Sci Technol 11:843–872

Marsden ID, Rainbow PS (2004) Does the accumulation of trace metals in crustaeceans affect their ecology-the amphipod example? J Exp Mar Biol Ecol 300:373–408

Martin JH (1990) Glacial–interglacial CO_2 change: the iron hypothesis. Paleoceanography 5:1–13

Marx SK, McGowan HA (2010) Long-distance transport of urban and industrial metals and their incorporation into the environment: sources, transport pathways and historical trends. In: Zereini F, Wiseman CLS (eds) Urban airborne particulate matter: origin, chemistry, fate and health impacts. Springer, Berlin

McComb J, Alexander TC, Han FX, Tchounwou PB (2014) Understanding biogeochemical cycling of trace elements and heavy metals in estuarine ecosystems. J Bioremediat Biodegrad 5:1000e148. https://doi.org/10.4172/2155-6199.1000e148

McGeer JC, Szebedinszky C, McDonald DG, Wood CM (2000) Effects of chronic sublethal exposure to waterborne Cu, Cd, or Zn in rainbow trout 1: ionoregulatory disturbance and metabolic costs. Aquat Toxicol 50:233–245

McLusky DS, Bryant V, Campbell R (1986) The effects of temperature and salinity on the toxicity of heavy metals to marine and estuarine invertebrates. Oceanogr Mar Biol-Ann Rev 24:481–520

Mead C, Herckes P, Majestic BJ, Anbar AD (2013) Source apportionment of aerosol iron in the marine environment using iron isotope analysis. Geophys Res Lett 40:5722–5727. https://doi.org/10.1002/2013GL057713

Millero FJ, Woosley R, DiTrolio BR, Waters J (2009) Effects of the ocean acidification on the speciation of metals in seawater. Oceanography 22:72–85

Mills MM, Ridame C, Davey M, La Roche J, Geider RJ (2004) Iron and phosphorous co-limit nitrogen fixation in the eastern topical North Atlantic. Nature 429:292–232

Mondal P, Reichelt-Brushett AJ, Jonathan MP, Sujitha SB, Sarkar SK (2017) Pollution evaluation of total and acid-leachable trace elements in surface sediments of Hooghly River Estuary and Sundarban Mangrove Wetland (India). Environ Sci Pollut. https://doi.org/10.1007/s11356-017-0915-0

Mondal K, Ghosh S, Haque S (2018) A review on contamination, bioaccumulation and toxic effect of cadmium, mercury and lead on freshwater fishes. Int J Zool Stud 3(2):153–159

Moore JW (1991) Manganese. In: Inorganic contaminants of surface water. Springer series on environmental management. Springer, New York, NY. https://doi.org/10.1007/978-1-4612-3004-5_15

Morel F, Reinfelder J, Roberts S, Chamberlain C, Lee J, Yee D (1994) Zinc and carbon co-limitation of marine phytoplankton. Nature 369:740–742

Mudd GM (2010) Global trends and environmental issues in nickel mining: sulphides versus laterites. Ore Geol Rev 38:9–26

Nair M, Joseph T, Balachandran KK, Nair KKC, Paimpillil JS (2003) Arsenic enrichment in estuarine sediments: Impact of iron and manganese mining. In: Ahmed MF, Ali MA, Adeel Z (eds) Fate of arsenic in the environment, pp 137–145

Namieśnik J, Rabajczyk A (2010) The speciation and physicochemical forms of metals in surface waters and sediments. Chem Speciat Bioavailab 22(1):1–24. https://doi.org/10.3184/095422910X12632119406391

Naser HA (2013) Assessment and management of heavy metal pollution in the marine environment of the Arabian Gulf: a review. Mar Pollut Bull 72:6–13

Natale G, Basso N, Ronco A (2000) Effect of Cr (VI) on early life stages of three species of hylid frogs (Amphibia, Anura) from South America. Environ Toxicol 15(5):509–512. https://doi.org/10.1002/1522-7278(2000)15:5%3c509:AID-TOX21%3e3.0.CO;2-S

Neely WB (1980) Chemicals in the environment: distribution, transport, fate, analysis. Marcel Dekker, New York, p 245

Negri AP, Heyward AJ (2001) Inhibition of coral fertilization and larval metamorphosis by Tributyltin and copper. Mar Environ Res 51:17–27

Negri AP, Smith LD, Webster NS, Heyward A (2002) Understanding ship grounding impacts on a coral reef: potential effects of antifoulant paint contamination on coral recruitment. Mar Pollut Bull 44:111–117

Nordberg GF, Fowler BA, Nordberg M, Friberg LT (2007) Handbook on the toxicology of metals, 3rd edn. Elsevier Inc

Obi E, Okafor C, Igwebe A, Ebenebe J, Johnson Afonne O, Ifediata F, Orisakwe OE, Nriagu JO, Basu N (2015) Elevated prenatal methylmercury exposure in Nigeria: evidence from maternal and cord blood. Chemosphere 119:485–489

Okoro HK, Fatoki OS, Adekola FA, Ximba BJ, Snyman RG (2011) Sources, environmental levels and toxicity of organotin in marine environment—a review. Asian J Chem 23(2):473–482

Okoro HK, Fatoki OS, Adekola FA, Ximba BJ, Snyman RG (2012) a review of sequential extraction procedures for heavy metals speciation in soil and sediments 1:181. https://doi.org/10.4172/scientificreports.181

Pastorok RA, Bilyard GR (1985) Effects of sewage pollution on coral reef communities. Mar Ecol Prog Ser 21:175–189

Patel P, Bielmyer-Fraser GK (2015) The influence of salinity and copper exposure on copper accumulation and physiological impairment in the sea anemone, *Exaiptasia Pallida*. Comparative biochemistry and physiology part C 168: 39–47

Pawlik-Skowrońska B, Kaczorowska R, Skowroński T (1997) The impact of inorganic tin on the planktonic cyanobacterium *synechocystis aquatilis*: the effect of pH and humic acid. Environ Pollut 97(1–2):65–69

Pekey H (2006) The distribution and sources of heavy metals in Izmit Bay surface sediments affected by a polluted stream. Mar Pollut Bull 52:1197–1208

Peters EC, Gassman NJ, Firman JC, Richmond RH, Power EA (1997) Ecotoxicology of tropical marine ecosystems. Environ Toxicol Chem 16:12–40

Philips DJH, Rainbow PS (1994) Biomonitoring of trace aquatic contaminants, 2nd edn. London, Chapman and Hall

Pinsino A, Matranga V, Roccheri MC (2012). Manganese: a new emerging contaminant in the environment. In: Srivastava J (ed) Environmental contamination. ISBN: 978-953-51- 0120-8, InTech. http://www.intechopen.com/books/environmental-contamination/manganesea-new-eme rging-contaminant-in-the-environment

Purcell TW, Peters JJ (1998) Sources of silver in the environment. Environ Toxicol Chem 17:539–546

Rainbow PS (1995) Biomonitoring of heavy metal availability in the marine environment. Mar Pollut Bull 31(4–12):183–192

Rainbow PS (1997) Trace metal accumulation in marine invertebrates: marine biology or marine chemistry. J Mar Biol Assoc U K 77:195–210

Raiswell R, Canfield DE (2012) The iron biogeochemical cycle past and present. Geocheml Persp 1(1)

Reichelt-Brushett AJ (2012) Risk assessment and ecotoxicology: limitations and recommendations for ocean disposal of mine waste in the coral triangle. Oceanography 25(4):40–51. https://doi. org/10.5670/oceanog.2012.66

Reichelt-Brushett AJ, Harrison PL (1999) The effect of copper, zinc and cadmium on fertilization success of gametes from *scleractinian* reef corals. Mar Pollut Bull 38:182–187

Reichelt-Brushett AJ, Hudspith M (2016) The effects of metals of emerging concern on the fertilization success of gametes of the tropical scleractinian coral *Platygyra daedalea*. Chemosphere 150:398–406

Reichelt-Brushett AJ, Stone J, Howe P, Thomas B, Clark M, Male Y, Nanlohy A, Butcher P (2017) Geochemistry and mercury contamination in receiving environments of artisanal mining wastes and identified concerns for food safety. Environ Res 152:407–418

Richir J, Gobert S (2016) Trace elements in marine environments: occurrence, threats and monitoring with special focus on the coastal mediterranean. J Environ Anal Toxicol 6:349. https://doi. org/10.4172/2161-0525.1000349

Rieuwerts J (2015) The elements of environmental pollution, 1st edn. Routledge

Rodrigo AG (1989) Surficial sediment heavy metal association in the Avon-Heathcote estuary, New Zealand. N Z J Mar Freshw Resour: 263–274

Rodrigues P, Ferrari RG, dos Santos LN, Junior CAC (2019) Mercury in aquatic fauna contamination: a systematic review on its dynamics and potential health risks. J Environ Sci 84:205–218. https://doi.org/10.1016/j.jes.2019.02.018

Röllin H (2011) Manganese: environmental pollution and health effects. In:Nriagu JO (ed) Encyclopedia of environmental health

Selin NE (2009) Global biogeochemical cycling of mercury: a review. Annu Rev Environ Resour 34(1):43–63. https://doi.org/10.1146/annurev.environ.051308.084314

Siddiqui S, Bielmyer-Fraser GK (2015) Responses of the sea anemone, *Exaiptasia pallida*, to ocean acidification conditions and copper exposure. Aquat Toxicol 167:228–239

Siddiqui S, Goddard RH, Bielmyer-Fraser GK (2015) Comparative effects of dissolved copper and copper oxide nanoparticle exposure to the sea anemone, *Exaiptasia pallida*. Aquat Toxicol 160:205–213

Shah S (2008) Study of heavy metal accumulation in *scleractinian* corals of Viti Levu, Fiji Islands, unpublished MSc thesis. Faculty of Science, Technology and Environment, The University of the South Pacific, Fiji

Slemr F, Weigelt A, Ebinghaus R, Bieser J, Brenninkmeijer CAM, Rauthe-Schoch A, Hermann M, Martinsson BG, vanVelthoven P, Bönisch H, Neumaier M, Zahn A, Ziereis H (2018) Mercury distribution in the upper troposphere and lowermost stratosphere according to measurements by the IAGOS-CARIBIC observatory: 2014–2016. Atmos Chem Phys 18:12329–12343. https://doi. org/10.5194/acp-18-12329-2018

Smith LD, Negri AP, Philipp E, Webster NS, Heyward AJ (2003) The effects of antifoulant paint contaminated sediments on coral recruits and branchlets. Mar Biol 143:651–657

Solomon F (2008) Impacts of heavy metals on aquatic systems and human health. Mining.com. www.infomine.com/library/publications/docs/mining.com/Apr2008c.pdf

Solomon F (2009) Impacts of copper on aquatic ecosystems and human health. Mining.com. http://www.ushydrotech.com/files/6714/1409/9604/Impacts_of_Copper_on_Aquatic_Ecosystems_and_human_Health.pdf

Summer K, Reichelt-Brushett A, Howe P (2019) Toxicity of manganese to various lfe stages of selected marine cnidarian species. Ecotoxicol Environ Saf 167:83–94. https://doi.org/10.1016/j.ecoenv.2018.09.116

Suvarapu LN, Seo YK, Bael SO (2013) Speciation and determination of mercury by various analytical techniques. Rev Anal Chem 32(3):225–245. https://doi.org/10.1515/revac-2013-0003

Tagliabue A, Bowie AR, Boyd PW, Buck KN, Johnson KS, Saito MA (2017) The integral role of iron in ocean biogeochemistry. Nature 543:51–59

Takahashi Y, Furukawa T, Kanai Y, Uematsu M, Zheng G, Marcus MA (2013) Seasonal changes in Fe species and soluble Fe concentration in the atmosphere in the Northwest Pacific region based on the analysis of aerosols collected in Tsukuba, Japan. Atmos Chem Phys 13(7695–7710)

Tamás MJ, Sharma SK, Ibstedt S, Jacobson T, Christen P (2014) Heavy metals and metalloids as a cause for protein misfolding and aggregation. Biomolecules 4:252–267. https://doi.org/10.3390/biom4010252

Tchounwou PB, Yedjou CG, Patlolla AK, Sutton DJ (2012) Heavy metals toxicity and the environment. EXS 101:133–164. https://doi.org/10.1007/978-3-7643-8340-4_6

Tornero V, Hanke G (2016) Chemical contaminants entering the marine environment from sea-based sources: a review with a focus on European sea. Mar Pollut Bull 112:17–38

United States Environmental Protection Agency (1993) Standard methods for the examination of water and wastewater. Am Public Health Assoc, US

VanBriesen JM, Small M, Weber C, Wilson J (2010) Modelling chemical speciation: thermodynamics, kinetics and uncertainty. In: Hanrahan G (ed) Modelling of pollutants in complex environmental systems, Chap 4, vol II. ILM Publications, pp 133–149

Violante A, Cozzolino V, Perelomov L, Caporale AG, Pigna M (2010) Mobility and bioavailability of heavy metals and metalloids in soil environments. J Soil Sci Plant Nutr 10(3):268–292

Vosyliene MZ, Jankaite A (2006) Effect of heavy metal model mixture on rainbow trout biological parameters. Ekologia 4:12–17

Wang WC, Mao H, Ma, Yang WX (2014a) Characteristics, functions and applications of metallothioneins in aquatic vertebrates: review article. Front Mar Sci 1(Article 34):1–12. https://doi.org/10.3389/fmars.2014.00034

Wang Z, Kwok KWH, Lui GCS, Zhou GJ, Lee JS, Lam MHW, Leung KMY (2014b) The difference between temperate and tropical saltwater species' acute sensitivity to chemicals is relatively small. Chemosphere 105:31–43. https://doi.org/10.1016/j.chemosphere.2013.10.066

Wang R, Balkanski Y, Boucher O, Bopp L, Chappell A, Ciais P, Hauglustaine D, Penuelas J, Tao S (2015) Sources, transport and deposition of iron in the global atmosphere. Atmosp Chem Phys 15:6247–6270. www.atmos-chem-phys.net/15/6247/2015/. https://doi.org/10.5194/acp-15-6247-2015

Weigelt A, Slemr F, Ebinghaus R, Pirrone N, Bieser J, Bödewadt J, Esposito G, van Velthoven PFJ (2016) Mercury emissions of a coal-fired power plant in Germany. Atmosp Chem Phys 16:13653–13668. www.atmos-chem-phys.net/16/13653/2016/. https://doi.org/10.5194/acp-16-13653-2016

Wells ML, Priceb NM, Bruland KW (1995) Iron chemistry in seawater and its relationship to phytoplankton: a workshop report. Mar Chem 48:157–182

World Health Organization (2017) Lead poisoning and health. http://www.who.int/mediacentre/factsheets/fs379/en/

World Health Organization (1996) Trace elements in human nutrition and health. WHO Library Cataloguing, Geneva, p 361

Wu X, Cobbina SJ, Mao G, Xu H, Zhang Z, Yang L (2016) A review of toxicity and mechanisms of individual and mixtures of heavy metals in the environment. Environ Sci Pollut Res 23:8244–8259. https://doi.org/10.1007/s11356-016-6333-x

Wuana RA, Okieimen FE (2011) Heavy metals in contaminated soils: a review of sources, chemistry, risks and best available strategies for remediation. In: International Scholarly Research Network (ISRN) ecology, vol 2011, Article ID 402647, 20 pp. https://doi.org/10.5402/2011/402647

Ye BJ, Kim BG, Jeon MJ, Kim SY, Kim HC, Jang TW, Chae HJ, Choi WJ, Ha MH, Hong YS (2016) Evaluation of mercury exposure level, clinical diagnosis and treatment for mercury intoxication. Ann Occup Environ Med 28:5. https://doi.org/10.1186/s40557-015-0086-8

Zeng X, Chen X, Zhuang J (2015) The positive relationship between ocean acidification and pollution. Mar Pollut Bull 91:14–21. https://doi.org/10.1016/j.marpolbul2014.12.001

Chapter 2
Coral Reef Ecosystem

Abstract Coral polyps are the building blocks of coral reefs. Coral reef ecosystem consists of a high diversity of marine species yet is among the world's most fragile and endangered ecosystem. A highly complex mutualistic symbiosis exists between the coral polyp and the single celled algae (zooxanthellae) *Symbiodinium spp*, thus forming the trophic and structural foundation of coral reef ecosystem. Being photosynthetic organisms, zooxanthellae capture sunlight and turn it into energy-rich compounds, which is transferred to the cells of the polyp. In return, the zooxanthellae are nourished with nutrients such as nitrogen and phosphates and gets shelter. Corals get stressed when environmental conditions are altered and this leads to the expulsion of the symbiotic zooxanthellae. Coral reefs are beneficial as they provide a source of livelihood for people; stabilize coastlines and provide nursery homes for juvenile fish and shelter mangroves and wetlands from oceanic waves. Coral reefs degrade naturally as well as anthropogenically. Corals excrete dimethylsulponiopropionate (DMSP), which may have a profound impact on the climate system.

Keywords Coral polyps · Zooxanthallae · Mutualistic symbiosis · Oceanic acidification · DMSP

2.1 Coral Reefs—An Overview

Corals are either hermatypic (reef building) or ahermatypic (non-reef-building). Hermatypic (*scleractinian* or stony or hard) corals thrive in shallow water habitats and contain millions of zooxanthellae, and create the primary reef framework. Ahermatypic corals lack zooxanthellae but secrete complex aragonite exoskeletons, and usually do not contribute significantly to reef formation.

Coral reefs are ecologically complex ecosystems engineered primarily by stony corals that support hundreds of thousands of species and valuable structural habitats (Fisher et al. 2015; Madin et al. 2016; Putnam et al. 2017); are amongst the world's most fragile and endangered ecosystems (Howard and Brown 1984; Bielmyer-Fraser et al. 2018); and yet consist of an enormous cultural, ecological and economic value (Mollica et al. 2018). Coral reef ecosystems thrive in low nutrient (such as inorganic phosphate, nitrate, nitrite and ammonia) waters and are extremely productive.

Corals from the phylum Cnidaria necessitate light for survival and growth, with salinities in the range 32–40 parts per thousand (‰) and seawater temperatures ranging from 18 to 30 °C (Veron 1986; Hoegh-Guldberg 1999; Souter and Lindén 2000). Most corals are colonial organisms, consisting of thousands of individual coral polyps (Souter and Lindén 2000), which secrete limestone/calcium carbonate ($CaCO_3$) that functions as the skeleton and over time builds the 3D-structure of the coral reef.

It is well documented that the calcium carbonate skeletons of corals record variations of the environmental conditions in which they live and these geochemical records have been used extensively in paleoclimatic and paleoceanographic research (Goreau 1977; Guzman and Jimenez 1992; Bastidas and Garcia 1999; Esslemont 1999; Fallon et al. 2002; David 2003; Runnalls and Coleman 2003; Prouty et al. 2008; Wei et al. 2009; McGregor et al. 2013; Zinke et al. 2014; Chen et al. 2015). For instance, Li et al. (2020) stated that the ability of corals to record upwelling, riverine runoff, and pollution from various sources depends upon the concentration of heavy metals in the coral aragonite matrix, which precisely reveals the concentrations of heavy metals in the marine environment.

2.2 Zooxanthellae in Corals

A symbiotic relationship exists between the coral polyp and the single celled algae (zooxanthellae) *Symbiodinium* spp; which live within the cells of the polyp (Muscatine and Porter 1977; Shah 2008; Pogoreutz et al. 2017; Bielmyer-Fraser et al. 2018) and are in extremely high densities (greater than 10^6 cm^{-2}) (Muscatine and Porter 1977; Allemand and Osborn 2019). This mutualistic symbiosis between the coral polyp and *Symbiodinium* spp is highly complex (Belda-Baillie et al. 2002) and forms both the trophic and structural foundation of coral reef ecosystems (Muscatine 1990; Ganot et al. 2011; Hume et al. 2015). The success of this symbiosis is evident in the geological persistence of coral reefs of more than 200 million years (Veron 1995) and the geographical area occupied by modern coral reefs in tropical and subtropical waters of over 280,000 km^2 (Spalding et al. 2001).

The photosynthetic symbionts (zooxanthallae) fix large quantities of carbon dioxide thus producing reduced organic carbon which is translocated to the host as mobile compounds (photosynthates) such as glycerol, glucose, lipids and amino acids (Venn et al. 2008; Ganot et al. 2011; Tremblay et al. 2012; Muller-Parker et al. 2015). These compounds provide the coral with energy for respiration, growth, reproduction and the deposition of its calcium carbonate skeleton (Muscatine et al. 1983; Muscatine 1990; Lesser 2004; Tremblay et al. 2012).

In return, the zooxanthellae are nourished with nutrients such as inorganic nitrogen, phosphorus and inorganic carbon (Venn et al. 2008; Ganot et al. 2011) and gets shelter. The coral-zooxanthellae symbiosis is very sensitive to increases in temperature however; and changes of as little as 1 °C above the average summer maximum can lead to a breakdown of the symbiosis (Mieog et al. 2009) and thus the loss of symbiotic algae from the host (Muller-Parker et al. 2015).

The photosynthetic pigments of the symbiotic algae are responsible for the color of the coral (Jones 1997; Falkowski and Dubinsky 1981; Dove et al. 2001; Fine and Loya 2002) and this can vary from white, yellow, brown and olive to red, green, blue and purple (Holden 1999).

Corals stress out when environmental conditions are altered and this leads to the expulsion of the *Symbiodinium* spp (or a decrease in the photosynthetic pigment concentration within the zooxanthellae) (Holden 1999). Environmental perturbations (such as an increase in seawater temperature) lead to global climate change and induce dysfunction and collapse of symbiosis leading to zooxanthellae loss or so-called 'bleaching'; which causes visible paling of coral colonies, and this phenomenon has led to severe worldwide decline of coral reefs (Brown and Howard 1985; Guzman and Jimenez 1992; Gardener et al. 2003; Baker et al. 2008; Weis 2008; Bielmyer et al. 2010; Ganot et al. 2011; Bielmyer-Fraser et al. 2018).

'Bleaching' refers to the change of coloration, revealing the white skeleton, and is considered the initial response to a stress of the host and/or the photosynthetic symbiont (Jones 1997; Perez et al. 2001; Baker et al. 2008; Muller-Parker et al. 2015). The extent of bleaching depends on the type of stress and the duration of it. Corals may recover from bleaching, depending on the intensity and duration of the stress, but if the algal symbiont communities are not restored relatively quickly (typically 2–4 weeks); corals may die (Mieog et al. 2009; Bielmyer et al. 2010; Bielmyer-Fraser et al. 2018).

2.3 Coral Nutrition

Reef building corals exhibit mixotrophy, relying on both the photoautotrophic products of their endosymbiotic algae and the nutrients acquired through heterotrophic predation (Goreau et al. 1971; Muscatine and Porter 1977; Hughes and Grottoli 2013). Mixotrophy results in a complex cycling of inorganic and organic carbon between the coral host, the skeleton it secretes, and its endosymbiotic algae (Reynaud et al. 2002; Hughes et al. 2010). Burmester et al. (2018) reported that when obtaining energy via photosynthesis, the coral holobiont (host animal plus symbionts) is functioning as an autotroph, and when obtaining energy via predation, it is functioning as a heterotroph.

Autotrophic (or photoautotrophic) nutrition is essentially through the mutualistic symbiosis (mentioned in Sect. 2.2) between the coral polyp and the *Symbiodinium spp* (zooxanthellae) and this nutritional interaction is one of the key reasons for the success of corals in nutrient poor tropical waters (Tremblay et al. 2012). The zooxanthellae play an important role in the energy budget of hermatypic corals by capturing sunlight and converting into energy-rich compounds (photosynthates); which are then transferred to the cells of the coral polyp and in exchange get shelter and nourishment (with nutrients such as nitrogen and phosphates). The translocated photosynthates can provide up to 100% energy to the coral host (Muscatine 1990; Muscatine et al. 1981; Falkowski et al. 1984; Tremblay et al. 2012).

Heterotrophy plays an important role in the energy budget of scleractinian corals since many nutrients cannot be supplied by photosynthesis and have to be acquired through feeding (Houlbréque and Ferrier-Pagés 2009; Tagliafico et al. 2018). Heterotrophic nutrition is considered most important at night (Porter 1976; Muscatine and Porter 1977) and this occurs when the sessile coral colonies fulfill a role as natural filter and water cleaning systems by capturing plankton, sediments, seston and organic particles from the water column (Seeman 2013). Corals can ingest (meso/macro) sized zooplankton (Grottoli et al. 2006; Palardy et al. 2008; Anthony et al. 2009; Grottoli et al. 2014; Levas et al. 2015) by nematocyst discharges and tentacle grabbing (reviewed by Houlbréque and Ferrier-Pagés 2009); pico- and nano-plankton (Tremblay et al. 2012); suspended particulate matter (SPM) (Anthony 1999a, b; Anthony and Fabricius 2000; Bessell-Browne et al. 2017). The SPM on coral reefs consists of diverse food sources such as detrital matter, resuspended sediment, coral mucus and excretory products from other animals and these are used through heterotrophic feeding.

Heterotrophy increases in high turbidity and low light circumstances (Anthony 2000; Anthony and Fabricius 2000); and following bleaching events (Grottoli et al. 2006; Bessell-Browne et al. 2014). Increase in habitat depth leads to a decrease in photosynthetic production, due to light limitation and thus some coral species depend on zooplankton grazing to sustain metabolism (Palardy et al. 2008; Tremblay et al. 2012). With increasing depth, the light availability reduces due to attenuation by the water column (Williams et al. 2018) and the heterotrophic energetic subsidies such as zooplankton and particulate organic matter increases (Hamner et al. 1988; Genin et al. 2005; Palardy et al. 2008).

Furthermore, heterotrophic nutrition dominates during bleaching events owing to a decrease in photosynthetic production (due to loss of symbionts or photosynthetic pigments) (Hughes et al. 2010; Tremblay et al. 2012). Elevated sea surface temperatures (SSTs) enable corals to lose their endosymbiotic algae and/or endosymbiotic algal pigments rendering them pale white (Hoegh-Guldberg and Smith 1989, Jokiel and Coles 1990; Glynn 1996; Brown 1997a, b; Hoegh-Guldberg 1999; D'Croz et al. 2001; Rodrigues and Grottoli 2007). Therefore, there is a dramatic decrease in photosynthesis and hence a deficiency in coral carbon budget of up to 80% (Muscatine et al. 1981; Falkowski et al. 1993; Grottoli et al. 2006; Palardy et al. 2008; Tremblay et al. 2012; Grottoli et al. 2014; Levas et al. 2015).

2.4 Benefits of Coral Reefs

Coral reefs form a unique ecosystem with a high biodiversity, are exceptionally valuable and thrive in nutrient poor waters. Coral reefs constitute only 0.2% of the world's marine ecosystem and estimated to harbour around one third of all described marine species (Reaka-Kudla 1997; Bryant et al. 1998; Reaka-Kudla 2001). World-wide approximately 3 billion people depend more or less directly on coral reefs for a significant part of their livelihood, obtaining their protein needs or other essential commodities (Bryant et al. 1998; Burke et al. 2011; Teh et al. 2013; Ferrario et al. 2014).

Allemand and Osborn (2019) reported that coral reefs provide important 'ecosystem services' that generate the conditions for human communities to settle and potentially thrive in coastal areas adjacent to the reefs. The 'ecosystem services' generally provided include reef tourism (income generation), coastal protection, primary productivity, food production for the local population, cultural services and supporting biodiversity. Albright and Cooley (2019) described that many people have cultural and emotional connections to reefs, thus increasing the overall importance of healthy coral reef systems to humans.

Coral reefs consist of a diversity of marine organisms such as crustaceans, fish, molluscs, algae and turtles, which are a food source for humans (UNEP/IUCN 1988; Costanza et al. 1997; Allemand and Osborn 2019). Coral reefs also provide a source of livelihood for many people through the fishing and tourism industry (Souter and Lindén 2000; Allemand and Osborn 2019). Spalding et al. (2017) reported that the major source of income comes from tourism and recreational activities, where reef tourism accounts for 30% of global reefs and 9% of coastal tourism worldwide. Reef tourism is particularly important for the economies of the Small Island Developing States (SIDS) (Allemand and Osborn 2019); as it becomes the major income earner for SIDS.

Coral reefs stabilize coastlines, thus protecting the shoreline from erosion by oceanic swells and tropical storms (Hoegh-Guldberg 1999; Souter and Lindén 2000; Johnson and Marshall 2007; Albright and Cooley 2019; Allemand and Osborn 2019); acting as an important natural barrier of storms and floods and protecting about 71,000 km of coastline worldwide (Beck et al. 2018). Coral reefs protect coasts by mitigating wave energy and storm surges (Albright and Cooley 2019) and hence reduce wave energy by up to 97% and the wave height by 84% (Ferrario et al. 2014). Coral reefs also help in the generation of sand (Perry et al. 2015), which is closely linked to reef tourism (Spalding et al. 2017). In fact, tourists prefer white sandy beaches and usually opt for tropical island countries for their vacations.

Coral reefs protect essential lagoon and mangrove habitat for vulnerable life stages of a wide range of commercial and non-commercial species (Johnson and Marshall 2007), together with providing vital spawning habitats for corals. The coral spawning environmental window is set as 5 days before spawning to 7 days after-wards (Jones et al. 2015) and involves the synchronous, multi-specific release of gametes by broadcast spawning coral species.

Furthermore, coral reefs contribute significantly to the well-being of man both now and in the future (Allemand and Osborn 2019). Coral reefs possess a diverse genetic reservoir with medicinal properties, and consequently get exploited in the search of new drugs, thus aiding humanity (Bruckner 2002; UNEP 2005; Cooper et al 2014).

Unique compounds obtained from coral reefs have helped in the treatment of cardiovascular diseases, ulcers, leukemia, and skin cancer and in the treatment for people infected with HIV (UNEP 2005). Corals possess anticarcinogenic properties (Ruiz-Torres et al. 2017) and anti-inflammatory, anti-cancer and anti-oxidant activities (Wei et al. 2013).

2.5 Threats to Coral Reefs

Coral reefs are the most biodiverse and delicate ecosystems on the planet, yet are the most susceptible to stress. The stressors could be acute or chronic, direct or indirect, and natural or anthropogenic. Highlighted below are some common threats, which cause coral reef degradation globally.

2.5.1 Natural Causes of Coral Reef Degradation

Local as well as global factors induce coral reef degradation naturally. Some examples include natural disasters (e.g. hurricanes/tropical storms/typhoons, earthquakes) (Rogers 1993; El-Naggar 2020); El Nino Southern Oscillation (ENSO) events and global warming (leading to bleaching events); coral diseases and crown of-thorns (COTS) starfish outbreaks (El-Naggar 2020); tidal fluctuations (Fadlallah et al. 1995); and high/low temperature extremes, freshwater plumes after heavy downpour and runoff events (McCulloch et al. 2003).

Tropical storms, typhoons or hurricanes do a great deal of damage such as breaking corals by strong wave action. Though massive, the damage tends to be localized and restricted to a particular area only (Souter and Lindén 2000). Stronger storms enhance the intensity of floods (Anthony 2016) which lead to sedimentation issues in the marine environment. Sedimentation causes turbid environments on the coral reef ecosystems. The intensity and frequency of tropical cyclones is more severe and enhanced for small island countries.

The corallivorous starfish, COTS (*Acanthaster planci*) become a common natural threat to corals when they occur in elevated numbers as they generally feed on coral tissue, thus damaging hard cover corals (Souter and Lindén 2000). If the damage is not severe, then recovery of the coral reef from the outbreaks of COTS may take 20–40 years (El-Naggar 2020). *Acanthaster planci* can produce many million babies during 1 year, however, with an increase in the nutrients from sewage and overharvesting of its predator, its population can escalate (El-Naggar 2020).

Acanthaster Planci greatly reduces coral cover and feeds preferentially on several coral species, which are highly susceptible to bleaching (Baker et al. 2008).

Furthermore, white and black band disease as well as bioeroders leads to the degradation of coral reef ecosystems to some extent. When under stress, coral reefs suffer many bacterial infections. This is due to production of protective mucus, which promotes the growth of many blue green algae responsible for black band disease (El-Naggar 2020). Excessive production of mucus is attributed to natural and anthropogenic influences such as global warming, toxic chemicals, pollution, eutrophication and increased sedimentation (El-Naggar 2020). Coral bleaching or heat stress expedites coral diseases.

Endolithic borers such as photosynthetic algae, heterotrophic fungi and bacteria, sponges, polychaete worms and mollusks cause bioerosion, which is the biological breakdown of limestone skeleton and reef frameworks. Baker et al. (2008) highlighted that disturbance such as eutrophication, sedimentation, epizootics and bleaching cause significant declines in coral cover and result in a loss of reef structures.

Bleaching can be triggered by high sea surface temperatures (SSTs), which are associated with climate change; solar radiation and El-Nino events (Douglas 2003), and bacterial infection as well as contamination from chemicals and pesticides (Souter and Lindén 2000). Bleaching is associated with high irradiance environments experiencing unusually warm conditions (1.0–1.5 °C above seasonal maximum mean temperatures) (Baker et al. 2008) and thus this may impair the symbiosis between the coral and its zooxanthallae, causing its breakdown (Putnam et al. 2017; Weijerman et al. 2018; Allemand and Osborn 2019).

It is well documented that increased solar energy [both in visible (400–700 nm) and ultraviolet region (290–400 nm) regions of the electromagnetic spectrum] have been associated in mass coral bleaching (Hoegh-Guldberg and Smith 1989; Brown et al. 1994; Fitt and Warner 1995; Shick et al. 1996; Brown 1997a, b; Lesser 1997, Brown et al. 1999; Dunne and Brown 2001; Coles and Brown 2003; Lesser 2004; Smith et al. 2005).

With a steady increase of ocean temperatures, many shallow water, tropical, reef-building corals live in excess of their upper thermal tolerance limits (Fitt et al. 2001), which varies geographically. Thermal anomalies repeatedly lead to the breakdown of the symbiotic relationship of corals with the *Symbiodinium spp*. Global coral bleaching events have intensified and are happening at greater frequencies over the last two decades and therefore coral reef communities have less recovery time, leading to widespread losses of coral cover and species diversity from tropical coral reef systems (Hughes et al. 2017, 2018).

Localized bleaching normally occurs and results from factors such as chemical spills, sedimentation, ship grounding and decrease in oceanic salinity by heavy rains or even from flood plumes (UNEP 2005). For instance, bleaching can be induced by cyanide exposure resulting in symbiont expulsion (Jones et al. 1999).

2.5.2 Anthropogenic Causes of Coral Reef Degradation

Coral reefs are exposed to anthropogenic disturbances which can range from local (e.g. overfishing, coastal development, changes in water quality and pollution) to global in scope (e.g. climate change, ocean acidification and hypoxia) (Pörtner et al. 2005; Hoegh-Guldberg et al. 2011, 2017; Putnam et al. 2017). Activities such as deforestation, agricultural pollution, dredging and mining operations contribute to the worsening quality of reef waters (Biscéré et al. 2018). It is remarkable to mention that numerous of the ecological changes in coral reefs could be traced back to anthropological activities that began centuries ago with recent rates of change escalating and leading to unprecedented loses of ecosystem services (Hoegh-Guldberg et al. 2019).

2.5.2.1   Local Disturbances

Local disturbances include destructive fishing with dynamite or cyanide; unsustainable tourism; pollution related to fertilizers and pesticides from agriculture and the evacuation of wastewater; sedimentation; proliferation of *Acanthaster* starfish; and the use of coral skeletons as a building material (Burke et al. 2011); overexploitation of marine resources; careless anchoring of boats (Hoegh-Guldberg et al. 2011); recreational damage; and excess access to corals for research purposes.

Coral reefs are increasingly exposed to growing loads of nutrients, sediments, pollutants discharged from land (Fabricius 2005) and metal contamination (Marques et al. 2019). Such disturbances can result in significant loss of ecosystem services impacting on social, cultural, biological and economic values (Bainbridge et al. 2018). Anthropogenic inputs can cause changes to the water quality and thus disrupt key chemical interactions among reef organisms (Peters et al. 1997; Bielmyer et al. 2010).

Anthropogenically induced stress to coral reefs is mainly associated with high population densities in coastal areas and the combined exposure of multiple interacting stressors (Bielmyer-Fraser et al. 2018; Maximillian et al. 2019). The expansion of human population in coastal areas and the accompanying urbanization and agricultural activities have increased the discharge of industrial wastes and untreated sewages into adjacent coral reefs (Hédouin et al. 2011) and these are escalating at an alarming rate.

The presence of humans within the proximity of coral reefs usually results in an elevated input of nutrients into reef waters (D'Angelo and Wiedenmann 2014). Municipal and industrial effluents add nitrogen, phosphorous, and industrial chemicals to the coastal environment, leading to eutrophication that produces a degradation of the marine food web (Maximillian et al. 2019); and at the same time leading to a reduction in the amount of light reaching the zooxanthellae and hence reduces photosynthesis, which in turn slows the coral growth and this may eventually lead to its death.

Activities such as coastal construction of breakwaters and marinas; land reclamation; beach nourishment; construction of ports and aquaculture facilities are carried out to meet the growing economic and societal demands in the coastal zone worldwide (Erftemeijer et al. 2012; Maximillian et al. 2019). Infrastructural and extension of agricultural developments to marginal lands have often led to increased turbidity and sedimentation in the surrounding coastal areas.

Natural resuspension events, terrestrial runoff, dredging and dredging related activities (such as dredge material disposal in offshore disposal grounds) can temporarily increase suspended sediment concentrations in the water column (Jones et al. 2015, Duckworth et al. 2017) and once suspended, fine terrestrial clays have the potential to remain in suspension for long periods (Storlazzi et al. 2015). Prouty et al. (2010) mentioned that terrestrial runoff contains sediment, nutrients as well as pollutants that can affect the health and resilience of corals.

Erftemeijer et al. (2012) reported that dredging operations have contributed to the loss of coral reef habitats, directly due to the removal or burial of reefs, and indirectly as a consequence of lethal or sublethal stress to corals caused by elevated turbidity and sedimentation. Collective effects linked with sediment inputs include direct mortality, reduced growth, lower calcification rates and reduced productivity, bleaching, increased susceptibility to diseases, physical damage to coral tissue and reef structures (breaking, abrasion), and reduced regeneration from tissue damage (Erftemeijer et al. 2012).

Enhanced sediment load is attributed to human activities, which is considered a major threat to coral reefs (Bryant et al. 1998; Hughes et al. 2003; Wilkinson 2004). High turbidity and sedimentation rates may depress coral growth and survival due to attenuation of light available to symbiotic zooxanthallae and redirection of energy expenditures for clearance of settling sediments (Erftemeijer et al. 2012). Sediments also affect the fertilization of broadcast spawning coral species (Ricardo et al. 2018).

Furthermore, sedimentation can affect coral recruitment and have impacts on other (non-coral) reef-dwelling organisms, reduced species diversity and increased mortality, reduced calcification and growth rates, decreased net production and increased respiration (Rogers 1990; Crabbe and smith 2005; Fabricius 2005; Jupiter et al. 2008; Erftemeijer et al. 2012).

Chemical pollution, associated with run off, is one of the most wide-spread anthropogenic disturbances to the marine environment and usually includes pernicious pollutants such as polychlorinated biphenyls (PCBs), pesticides and trace elements (Burke et al. 2011; Fey et al. 2019). Agricultural, industrial and urban activities contribute to chemical pollution. Coastal agricultural practices have led to an increase in pesticides and fungicides in waterways. Run-off from agricultural activities increases pollution especially in the form of nitrogen and phosphate (De-Bashan and Bashan 2004) thus causing eutrophication. Eutrophication is simply the gradual increase in the concentrations of nutrients in a water body; usually from runoff from land and these lead to excessive growth of algae in the water body. Pesticides have been identified as important non-point source pollutants since residues have been detected in several coral reef systems (Haynes et al. 2000; Lewis et al. 2009; Brodie et al. 2012).

Fonseca et al. (2019) highlighted that metals turn out to be one of the main pollutants in the marine environment and a major threat to the growth and reproduction of stony corals and especially Cu contamination can eventually cause ecological and economic disservice associated with the loss of integrity of coral reefs. Negri et al. (2011) mentioned that pollution and increased SST together impact upon corals during the vulnerable early stages of their development.

2.5.2.2 Global Disturbances

Local threats weaken corals, thus reducing their resilience to global stresses (Carilli et al. 2009).

Global climate change and disease, ocean acidification, habitat destruction, pollution from coastal development and anthropogenic threats, deteriorating water quality in coastal environment related to land use changes in the catchment basins, and heavily populated areas have been associated with the global decline of the coral reef ecosystems (Hoegh-Guldberg 1999; Hughes et al. 2003; Pandolfi et al. 2003; Prouty et al.2008; Ateweberhan et al. 2013; Bielmyer-Fraser et al. 2018). The coexistence of these disturbances together have led to the destruction of coral reef ecosystems.

The following discusses how climate change in association with other factors influences the coral reef ecosystems.

Climate Change Implications on Coral Reefs

Climate change relates to the change in the state of the climate, identified by changes in the mean and/or variability of its properties and that, which persists for an extended period. The change in climate over time could be due to natural variability or as a result of human activities. Greenhouse gases (GHGs), namely carbon dioxide, methane, and nitrous oxide are the driving force of climate change. These gases allow some of the heat from the sun to be absorbed by the land and ocean, thus keeping the earth within a stable temperature range comfortable for plants and animals, a phenomenon called the greenhouse effect. Greenhouse effect is important for the existence of life on earth; for without it the earth's global temperature would be $-18\,°C$.

Ever since the industrial revolution, human activities have increased the concentration of GHGs in the atmosphere (IPCC 2007, 2013). Increased concentrations of GHGs cause climate change by trapping heat reflected from the land and ocean and preventing it from leaving the Earth's atmosphere. The trapped heat makes air and water temperatures warmer than usual and hence leads to global climate change. EPA (2016) reported that average seawater temperatures have risen at a rate of $0.07\,°C$ per decade over the past century.

The ocean, being the most important component of the climate system, is intrinsically linked to the atmosphere via gas exchange, heat transfer and biogeochemical cycles. This warmer air and ocean surface temperature has fatal impact on corals. It contributes to coral bleaching events and alters the ocean chemistry. Increase in ocean surface temperatures causes zooxanthallae to be expelled from corals, making them turn white. Marques et al. (2019) highlighted that tropical coral reefs are mostly vulnerable to climate change. Furthermore, global disturbances associated with climate change exacerbates acute and chronic local stressors from declining water quality and overexploitation of key species, to driving reefs towards the tipping point for functional collapse (Hoegh-Guldberg et al. 2007; Dalton et al. 2020).

Being a greenhouse gas, CO_2 has directly influenced the global temperature and also has been absorbed by the ocean where it is being consumed by autotrophic phytoplankton for photosynthesis. Once absorbed into the ocean, CO_2 undergoes a series of chemical reactions producing a weak acid; which creates a CO_2- carbonate equilibrium (discussed in the next section). An increase in the flux of CO_2 into the ocean waters as a result of increasing anthropogenic atmospheric CO_2 concentrations shifts this CO_2-carbonate balance leading to a phenomenon referred to as 'ocean acidification' (discussed in detail below).

Ocean Acidification

The release of carbon dioxide (CO_2) into the atmosphere by human activities leads to an increased flux of CO_2 into a mildly alkaline ocean, thus leading to an increased concentration of inorganic carbon and a reduction in pH, carbonate ion concentration, and the capacity of seawater to buffer changes in its chemistry (Wang et al. 2015; Yang et al. 2016); resulting in a phenomenon known as ocean acidification.

Ocean acidification results from the dissolution of carbon dioxide into oceanic waters subsequently forming carbonic acid (a weak acid); which dissociates to bicarbonate and further to carbonate ions (Bielmyer-Fraser et al. 2018) and releases hydrogen ions as shown in Eq. 2.1 below.

$$CO_{2(g)} + H_2O_{(1)} \rightarrow H^+_{(aq)} + HCO^-_{3(aq)} + 2H^+_{(aq)} + CO^-_{3(aq)} \qquad (2.1)$$

{Where HCO_3^- is the bicarbonate ion and CO_3^{2-} is the carbonate ion}

The release of hydrogen ions lowers the pH; which results in a reduction in the concentrations of both hydroxide (OH^-) and carbonate ions in most natural surface waters (Millero et al. 2009).

The dissolved forms of carbon dioxide reflect the pH of seawater and maintains it within relatively narrow limits by operating as a natural buffer to the addition of hydrogen ions, referred to as the 'carbonate buffer'. The carbonate buffer operates on the principle that if an acid (such as CO_2) is added to seawater, the additional hydrogen ions react with carbonate (CO_3^{2-}) ions and convert them to bicarbonate (HCO_3^-). This reduces the concentration of hydrogen ions (the acidity) such that the change in pH is much less than would otherwise be expected as shown by Eq. 2.2.

$$CO_{2(g)} + H_2O_{(1)} + CO^{2-}_{3(aq)} \rightarrow HCO^-_{3(aq)} + H^+_{(aq)} + CO_3(aq)^{2-} \rightarrow 2HCO^-_{3(aq)}$$
$$(2.2)$$

Hence, when atmospheric CO_2 dissolves in seawater, the oceans increase in acidity but, due to the carbonate buffer, the resultant solution is still slightly alkaline. The capacity of the buffer to restrict pH changes diminishes as increased amounts of CO_2 are absorbed by the oceans (The Royal Society 2005); thus causing the world's oceans to become warmer and more acidic (Weijerman et al. 2018).

The dissolved forms of carbon dioxide are important for the biological processes of marine organisms such as photosynthesis (the production of complex organic carbon molecules from sunlight) by marine algae; and calcification. Shifting the balance between these different carbon species may have dramatic effects on aquatic life, depending on which carbon species dominates (Bielmyer-Fraser et al. 2018).

The carbonate ion constitutes the essential brick for the formation of shells and skeletons of invertebrates, including the skeleton of corals (Allemand and Osborn 2019). Marangoni et al. (2019) stated that the tridimensional structure of coral reefs is built by calcifying organisms that are able to secrete carbonate skeletons through the biomineralization process. The carbonate skeletons are formed when carbonate ion reacts with the calcium ions to produce calcium carbonate, referred to as calcification, as shown by Eq. 2.3.

$$CO_{3(aq)}^{2-} + Ca_{(aq)}^{2+} \rightarrow CaCO_{3(s)} \tag{2.3}$$

Carbonate ion concentrations are often expressed relative to the saturation state of seawater with respect to aragonite, the principal crystal form of calcium carbonate crystals deposited by reef-building corals and many other marine calcifiers (Hoegh-Guldberg 2011). Changes in aragonite saturation state are considered a proxy for calcification rate (Langdon and Atkinson 2005).

Under acidic conditions, the concentration of carbonate ions reduces in seawater according to Eq. 2.4.

$$H_{(aq)}^{+} + CO_{3(aq)}^{2-} \rightarrow HCO_{3(aq)}^{-} \tag{2.4}$$

At lowered pH, corals cannot absorb the calcium carbonate that they need to maintain their skeletons. Orr et al. (2005) reported that as the carbonate becomes depleted over time; seawater can become undersaturated with respect to important calcium carbonate minerals that are fundamental building blocks for many important marine species, such as corals, zooplankton, and shellfish. The chemical modifications have profound biological impacts by modifying the quantities of carbonate necessary for the formation of the skeletons but especially by modifying the pH, a key parameter in physiology (Comeau et al. 2017).

Furthermore, ocean acidification can similarly affect other aspects of physiology of marine organisms including acid-base balance, energy metabolism, redox balance as well as behavior (Ishimatsu et al. 2008; Sokolova et al. 2015) and hence the possible interactions with pollutants such as metals.

Fluctuations in seawater chemistry as a result of ocean acidification can affect parameters such as solubility, speciation and distribution of metals in water and sediments, and thus affect the potential metal toxicity to marine organisms. The concentrations of the dissolved metals in the marine environment are typically low due to the relatively low solubility of trace metals in seawater and absorption of the metals on the sediments (Ivanina and Sokolova 2015).

Most metals are associated with organic matter; with a significant fraction in the form of metal-organic complexes (Zeng et al. 2015). Lowered pH affects the adsorption of metals to organic material as most organic particles in seawater are negatively charged (Millero et al. 2009) and hence surface sites become less available to adsorb metals (Crist et al. 1988; Wilde et al. 2006). Hence, the effect of pH on the speciation of metal-organic complexes in the marine environment is not as well characterized as the inorganic ligands due to their heterogeneous composition and the unknown structure of the organic ligands (Millero et al. 2009; Zeng et al. 2015).

Fluctuations in pH also affect the oxidation and reduction reaction rates of metals and their photochemical processes (Millero et al. 2009; Zeng et al. 2015). In the ocean surface waters, photochemical processes produce a number of free radicals that can change the oxidation state of a number of metals (Millero et al. 2009). Acidification usually increases the reduction rates more than the oxidation rates, as the latter is less pH dependent (Zeng et al. 2015). For example, the half-life of Fe (II) in seawater increases from 1 min to 24 min when pH decreases from 8.1 to 7.4 (Millero et al. 2009).

The speciation of metals changes due to ocean acidification and a shift in speciation leads to changes in the solubility, bioavailability and toxicity of those metals (Diaz and Rosenberg 2008; Millero et al. 2009; Marangoni et al. 2019).

A lowered pH causes a reduction in the concentrations of OH^- and CO_3^{2-}, which forms strong bonds in ocean water with divalent and trivalent cations (Zeng et al. 2015). Metals forming strong complexes with OH^- and CO_3^{2-} have a higher fraction in their free forms at lower pH (Millero et al. 2009) and thus ocean acidification could transform heavy metals into a specification more toxic to biota (Zeng et al. 2015).

Sediments become the ultimate reservoir for pollutants such as metals discharged into the marine environment and hence metals become unavailable. Metals in the sediment reside in different geochemical fractions, namely exchangeable and water soluble, carbonate-, organic-, Fe–Mn oxide-bound and residual and shows a high variation in changing factors such as pH, granulometry, temperature (Wang et al. 2015). Under reduced pH conditions, metals desorb from the sediments and organic ligands causing an elevated flux of dissolved metals into the water column (de Orte et al. 2014a, b) thus causing toxicity. Reactions such as adsorption/desorption, dissolution of carbonates, sulphides and iron oxyhydroxide occur at the sediment-water interface.

Hence, the coexistence of ocean acidification and pollution in many coastal regions may have combined effects on marine ecosystems in the forseeable future (Zeng et al. 2015). Bielmyer-Fraser et al. (2018) highlighted that both metal pollution and ocean acidification have shown to cause deleterious effects in aquatic organisms individually; however, the problem of dual exposure may be exacerbated because lower pH (increased acidification) causes changes in metal speciation, resulting in a shift to more toxic ionic metal species. Negri and Hoogenboom (2011), Fonseca et al. (2017) and Banc-Prandi and Fine (2019) deliberated that in the context of an era of climate change, the toxicity of metals is expected to be enhanced due to increasing seawater temperature and acidification, thus resulting in more severe impacts to coral reefs. Ocean acidification is a global phenomenon; with its impacts expressed at local levels (Lebrec et al. 2019).

Production and Fate of DMSP in *Scleractinian* Corals and Its Implications
in Climate Change

Corals (Van Alstyne et al.2008), polyps (Raina et al. 2013), phytoplankton species
confined to the classes Dinophyceae (dinoflagellates) and Prymnesiophyceae (Stefels
2000) and heterotrophic bacteria (Curson et al. 2017) are capable of producing
Dimethylsulponiopropionate (DMSP); which is a ubiquitous compound in the marine
sulphur cycle. Its production is attributed entirely to the activities of their algal
endosymbiont, *Symbiodinium* spp (Raina 2013) and is present in coral tissue,
symbiotic microalgae and coral mucus (Broadbent and Jones 2004).

The expulsion of coral symbionts and coral mucous to reef waters raises dimethylsulphide (DMS) and DMSP concentrations (Broadbent and Jones 2004) and this
makes coral reefs potential "hotspots" of atmospheric DMS production (Broadbent
et al. 2002; Broadbent and Jones 2004; Swan et al. 2012). Zhang et al. (2019) highlighted that the "hotspots" due to high DMS and DMSP coincide with high primary
productivity ocean areas. Stefels (2000) reported that algal biosynthesis of DMSP
occurs via assimilatory sulphate reduction, which is energy utilizing metabolic
process.

Marine phytoplankton provide their own antioxidant defense by utilizing the
sulphur substances DMSP, DMS, dimethylsulphoxide (DMSO), acrylic acid (AA)
and methanesulphinic acid (MSNA); which may function individually or simulta-
neously as an efficient antioxidant system to scavenge the harmful oxygen free-
radicals produced during elevated stress (Sunda et al. 2002). Jones and King (2015)
further highlighted that DMSP, DMS and acrylate also scavenge hydroxyl radicals
(OH) and produce DMSO. DMSO further reacts with hydroxyl radicals to produce
MSNA, which also scavenges hydroxyl radicals and other ROS (reactive oxygen
species) (Sunda et al.2002). The antioxidants (chemical defence compounds) protect
coral tissues from environmental stress, including that caused by high solar radiation
(Raina et al. 2013).

DMSP have many biological roles such as (a) osmoregulation in various species of
phytoplankton (Vairavamurthy et al. 1985; Stefels 2000); (b) an antioxidant response
in phytoplankton and coral (Sunda et al. 2002; Jones et al. 2007;Deschaseaux et al.
2014); (c) anti-predation (Otte and Morris 1994; Wolfe et al. 1997; Van Alstyne
et al. 2001; Van Alstyne and Houser 2003); (d) anti-bacterial activity (Sieburth 1960,
1961); (e) methyl donor in the synthesis of nitrogen based metabolites (Chillemi
et al. 1990); (f) chemo-attractant for a whole range of marine species (DeBose et al.
2008; Seymour et al. 2010; Knight 2012; Savoca and Nevitt 2014;)); (g) chemical cue
for bacteria (Seymour et al. 2010; Garren et al. 2013); (h) cyroprotection in polar
waters (Kirst et al. 1991; Nishiguchi and Somero 1992); (i) it is a metabolite of
many marine phytoplanktons as well as an overflow mechanism that allows cells to
maintain energy balance under sub-optimal conditions (Stefels 2000).

DMSP is the precursor of DMS, a strong smelling, volatile gas that is released
into the atmosphere and has a very short atmospheric lifetime of approximately
1–2 days. DMSP can be enzymatically cleaved by DMSP lyase to produce DMS,
acrylate and acrylic acid (Jones et al. 2014). DMS provides an estimated 28.1 Tg of

sulphur to the atmosphere via its sea-to-air exchange from global oceans (Lana et al. 2011). Oxidation of atmospheric DMS by hydroxyl radicals leads to the formation of sulphate aerosols, which act as cloud condensation nuclei (CNN), thus leading to greater cloudiness. Thus this can significantly affect solar radiation over the ocean (Charlson et al. 1987) and Raina et al. (2013) elucidate that cloud production, especially in the tropics, is an important regulator of climate because clouds shade the earth and reflect much of the sun's heat back into space.

Charlson et al. (1987) hypothesized that changes in phytoplankton activities can be mediated by changes in low level cloud cover and sunlight, thus regulating sea surface temperatures (SSTs). The hypothesis of Charlson et al. (1987) is referred to as the CLAW hypothesis (named after the first letter of each authors surname), which proposes that as temperatures increase due to GHG gas warming, greater amounts of DMS would be emitted into the atmosphere to produce more low level cloud cover, which would in turn lower sea surface temperatures, a type of ocean thermostat or climate feedback.

Production of DMSP was found to increase when corals are subjected to water temperatures that put them under heat stress (Raina et al. 2013) and thus elevated levels of DMS above stressed reefs (Jones et al. 2007). Hence, Deschaseaux et al. (2014) reported that DMSP was found to be a biomarker of stress in corals as their concentrations tend to increase as a response to environmental stress exposure in reef corals. However, declining coral numbers could lead to a decrease in the production of DMSP, thus this may impede cloud formation.

Corals are also exposed to environmental pollutants such as metals. Schwarz et al. (2013) reported that exposure to metals induces oxidative stress in organisms through cellular biochemical reactions that produce ROS including superoxide (O_2^-), hydrogen peroxide (H_2O_2), hydroxyl radical (•OH) and singlet oxygen ($^1O_2$). An increase in temperature and metal exposure increases the generation of ROS in aquatic organisms leading to the oxidation of biomolecules like lipids, proteins and nucleic acids (Monserrat et al. 2007).

In their study, Deschaseaux et al. (2018) found that DMSP concentrations significantly decreased in both the host and symbionts under the most elevated zinc treatment (1000 μg/L); suggesting that the decrease is a consequence of DMSP oxidation from increased levels of ROS. The reef-building coral *Acropora aspera* was used in the study of Deschaseaux et al. (2018) and it was found that extended exposure to high Zn concentrations in *A. aspera* induced a decrease in DMSP concentrations in the coral host and associated *Symbiodinium;* indicating that DMSP was consumed (or stopped being produced) in the coral holobiont under Zn contamination. In their study, Jones et al. (2007) expounded that raised levels of DMS above stressed reefs have led authors to suggest the involvement of DMSP is a corals stress response.

However, Yost et al. (2010) in his research reported that *Montastraea franksi* showed differential responses to Cu exposure; DMSP levels generally decreased with Cu dose in both the holobiont and the symbiont fractions whilst the DMSP levels exclusively increased in the symbiont at the highest Cu dose.

Fonseca et al (2019) reported that Cu contamination could cause ecological and economic disservice associated with the loss of integrity of coral reefs. In conjunction with other stressors such as ocean acidification, Cu leads to the activation of antioxidant system in sea anemones and corals (Siddiqui and Bielmyer-Fraser 2015; Bielmyer-Fraser et al. 2018). This occurs because Cu has been shown to cause the generation of ROS which leads to the degradation of macromolecules such as proteins, lipids and DNA (Chang et al. 1996; Luschak 2011). DMSP is important to coral reefs yet the variation on the DMSP levels could be dependent on the type of metal the corals are being exposed to.

2.6 Summary

Coral reef ecosystems thrive in nutrient poor waters and are highly diverse. The mutualistic symbiosis between the coral polyp and zooxanthallae helps form the trophic and structural foundation of coral reef ecosystems. Coral reefs substantiate to be beneficial in a variety of aspects such as providing nurseries to juvenile fish, abode for drug exploitation, stabilizing coastlines and a source of livelihood for many people.

However, coral reef ecosystems are under threat both naturally as well as anthropogenically. Under stress, due to altering environmental conditions, corals expel the symbiotic zooxanthallae, thus leading to bleaching. Pollutants, nutrients, sedimentation, and declining water quality adversely impacts corals leading to burial and smothering, decreased calcification rates and reduced recruitment; thus leading to the global decline of the coral reefs. Climate change also contributes to coral bleaching events and alters ocean chemistry. Oceanic acidification occurs when pH is lowered, thus corals cannot absorb the calcium carbonate that is needed to maintain their skeletons. Corals and polyps excrete DMSP naturally and production of DMSP increases when corals are subjected to higher water temperatures. DMSP is a precursor of DMS, which has a profound impact on the climate system.

Combined effects of multiple interacting stressors lead to the decline of coral reefs. In an era of climate change and ocean acidification, it becomes paramount to understand the biological effects of metal exposure to corals.

References

Albright R, Cooley S (2019) A review of interventions proposed to abate impacts of ocean acidification on coral reefs. Reg Stud Mar Sci 29: https://doi.org/10.1016/j.rsma.2019.100612

Allemand D, Osborn D (2019) Ocean acidification impacts on coral reefs: From sciences to solutions. Reg Stud Mar Sci 28: https://doi.org/10.1016/j.rsma.2019.100558

Anthony K (1999a) Coral suspension feeding on fine particulate matter. J Exp Biol Ecol 232:85–106

Anthony K (1999b) A tank system for studying benthic aquatic organisms at predictable levels of turbidity and sedimentation: case study examining coral growth. Limnol Oceanogr 44:1415–1422

Anthony K (2000) Enhanced particle feeding capacity of corals on turbid reefs (Great Barrier Reef, Australia). Coral Reefs 19:59–67. https://doi.org/10.1007/s003380050227

Anthony KRN (2016) Coral reefs under climate change and ocean: challenges and opportunities for management and policy. Annu Rev Environ Resour 41:59–81. https://doi.org/10.1146/annurev-environ-110615-085610

Anthony K, Fabricius K (2000) Shifting roles of heterotrophy and autotrophy in coral energetics under varying turbidity. J Exp Mar Biol Ecol 252:221–253

Anthony KRN, Hoogenboom MO, Maynard JA, Grottoli AG, Middlebrook R (2009) Energetics approach to predicting mortality risk from environmental stress: a case study of coral bleaching. Funct Ecol 23:539–550

Ateweberhan M, Feary DA, Keshavmurthy S, Chen A, Sheppard CRC (2013) Climate change impacts on coral reefs: synergies with local effects, possibilities for acclimation and management implications. Mar Pollut Bull 74:526–539

Bainbridge ZT, Lewis S, Bartley R, Fabricius K, Collier C, Waterhouse J, Garzon-Garcia A, Robson B, Burton J, Wenger A, Brodie J (2018) Fine sediment and particulate organic matter: a review and case study on ridge-to-reef transport, transformations, fates and impacts on marine ecosystems. Mar Pollut Bull 135. https://doi.org/10.1016/j.marpolbul.2018.08.002

Baker AC, Glynn PW, Riegl B (2008) Climate change and coral reef bleaching: An ecological assessment of long-term impacts, recovery trends and future outlook. Estuar Coast Shelf Sci 80(4):435–471. https://doi.org/10.1016/j.ecss.2008.09.003

Banc-Prandi G, Fine M (2019) Copper enrichment reduces thermal tolerance of the highly resistant Red Sea coral *Stylophora pistillata*. Coral Reefs. https://doi.org/10.1007/s00338-019-01774-z

Bastidas C, Garci'a E (1999) Metal content on the reef coral *Porites astreoides*: an evaluation of river influence and 35 years of chronology. Mar Pollut Bull 38:899–907

Beck MW, Losada IJ, Menendez P, Reguro BC, Diaz-Simal P, Fernandez F (2018) The global flood protection savings provided by coral reefs. Nat Commun 9:2186

Belda-Baillie CA, Baillie BK, Maruyama T (2002) Specificity of a model cnidarian dinoflagellate symbiosis. Biol Bull 202:74–85

Bessell-Browne P, Stat M, Thomson D, Clode PL (2014) *Coscinaraea marshae* corals that have survived prolonged bleaching exhibit signs of increased heterotrophic feeding. Coral Reefs 33:795–804. https://doi.org/10.1007/s00338-014-1156-z

Bessell-Browne P, Negri AP, Fisher R, Clode PL, Duckworth A, Jones R (2017) Impacts of turbidity on corals: the relative importance of light limitation and suspended sediments. Mar Pollut Bull. https://doi.org/10.1016/j.marpolbul.2017.01.050

Bielmyer GK, Grosell M, Bhagooli R, Baker AC, Langdon C, Gillette P, Capo TR (2010) Differential effects of copper on three species of *Scleractinian* corals and their algal symbionts (*Symbiodinium* spp). Aquat Toxicol 97:125–133

Bielmyer-Fraser GK, Patel P, Capo T, Grosell M (2018) Physiological responses of corals to ocean acidification and copper exposure. Mar Pollut Bull 133:781–790

Biscéré T, Ferrier-Pages C, Grover R, Gilbert A, Rottier C, Wright A, Payri C, Houlbrèque F (2018) Enhancement of coral calcification via the interplay of nickel and urease. Aquat Toxicol 200:247–256. https://doi.org/10.1016/j.aquatox.2018.05.013

Broadbent AD, Jones GB, Jones RJ (2002) DMSP in corals and benthic algae from the Great Barrier Reef. Estuar Coast Shelf Sci 55:547–555

Broadbent AD, Jones GB (2004) DMS and DMSP in mucus ropes, coral mucus, surface films and sediment pore waters from coral reefs in the Great Barrier Reef. Mar Freshw Res 55:849–855

Brodie JE, Kroon FJ, Schaffelke B, Wolanski EC, Lewis SE, Devlin MJ, Bohnet IC, Bainbridge ZT, Waterhouse J, Davis AM (2012) Terrestrial pollutant runoff to the Great Barrier Reef: an update of issues, priorities and management responses. Mar Pollut Bull 65:81–100. https://doi.org/10.1016/j.marpolbul.2011.12.012

Brown BE (1997a) Coral Bleaching: causes and consequences. Coral Reefs 16:129–S138

Brown BE (1997b) Adaptations of coral reefs to physical environmental stress. Adv Mar Biol 31:221–299

Brown BE, Howard LS (1985) Assessing the effects of 'stress' on corals. Adv Mar Biol 22:1–63

Brown BE, Le Tissier MDA, Dunne RP (1994) Tissue retraction in the scleractinian coral *Coeloseris mayeri*, its effect upon coral pigmentation and preliminary implications for heat balance. Mar Ecol Prog Ser 105:209–218

Brown BE, Ambarsari I, Warner ME, Fitt WK, Dunne RP, Gibb SW, Cummings DG (1999) Diurnal changes in photochemical efficiency and xanthophyll concentrations in shallow water reef corals: evidence for photoinhibition and photoprotection. Coral Reefs 18:99–105

Bruckner AW (2002) Life-saving products from coral reefs. Issues Sci Technol 18(3)

Bryant D, Burke L, McManus J, Spalding M (1998) Reefs at risk: a map-based indicator of threats to the world's coral reefs. World Resources Institute, Washington, DC, 56 pp

Burke L, Reytar K, Spalding M, Perry A (2011) Reefs at risk revisited. Washington DC, USA

Burmester EM, Breef-Pliz A, Lawrence NF, Kaufman L, Finnerty JR, Rotjan RD (2018) The impact of autotrophic versus heterotrophic nutritional pathways on colony health and wound recovery in corals. Ecol Evolut 8:10805–10816

Carilli JE, Norris RD, Black BA, Walsh SM, McField M (2009) Local stressors reduce coral resilience to bleaching. PLoS ONE 4(7):

Chang LW, Magos L, Suzuki T (eds) (1996) Toxicology of metals. Lewis/CRC Publishers, Boca Raton

Charlson RJ, Lovelock JE, Andrea MO, Warren SG (1987) Oceanic phytoplankton, atmospheric sulfur, cloud albedo and climate. Nature 326:655–661

Chen X, Wei G, Deng W, Liu Y, Sun Y, Zeng T, Xie L (2015) Decadal variations in trace metal concentrations on a coral reef: Evidence from a 159-year record of Mn, Cu, and V in a Porites coral from the northern South China Sea. J Geophys Res Oceans 120:405–416. https://doi.org/10.1002/2014JC010390

Chillemi R, Patti A, Morrone R, PIatelli M, Sciuto S (1990) The role of methylsulphonium compounds in the biosynthesis of N-methylated metabolites in Chondria coerulescens. J Nat Prod 53:87–93

Coles SL, Brown BE (2003) Coral bleaching -capacity for acclimatization and adaptation. Adv Mar Biol 46:183–223

Comeau S, Tambutté É, Carpenter RC, Edmunds PJ, Evensen NR, Allemand D, Ferrier-Pagés C, Tambutté S, Venn AA (2017) Coral calcifying fluid pH is modulated by seawater carbonate chemistry not solely seawater ph. Proc R Soc B Biol Sci 284 (1847):20161669, pii

Cooper EL, Hirabayashi K, Strychar KB, Sammarco PW (2014) Corals and their potential applications to integrative medicine. Evid-Based Compl Altern Med 184959:1–9

Costanza R, d'Arge R, de Groot R, Farber S, Grasso M, Hannon B, Limburg K, Naeem S, O'Neill RV, Paruelo J, Raskin RG, Sutton P, van den Belt M (1997) The value of the world's ecosystem services and natural capital. Nature 387:253–260

Crabbe MJC, Smith DJ (2005) Sediment impacts on growth rates of Acropora and porites corals from fringing reefs of Sulawesi, Indonesia. Coral Reefs 24:437–441. https://doi.org/10.1007/s00338-005-0004-6

Crist RH, Oberholser K, Schwartz D, Marzoff J. Ryder D, Crist DR (1988) Interactions of metals and protons with algae. Environ Sci Technol 22:755–760

Curson ARJ, Liu J, Bermejo Martínez A, Green RT, Chan Y, Carrión O, Williams BT, Zhang SH, Yang GP, Bulman Page PC, Zhang XH, Todd JD (2017) Dimethylsulfoniopropionate biosynthesis in marine bacteria and identification of the key gene in this process. Nat Microbiol 2: https://doi.org/10.1038/nmicrobiol.2017.9

Dalton SJ, Carroll AG, Sampayo E, Roff G, Harrison PL, Entwistle K, Huang Z, Salih A, Diamond SL (2020) Successive marine heatwaves cause disproportionate coral bleaching during fast phase transition from El Nino to La Nina. Sci Total Environ 715(136951). https://doi.org/10.1016/j.scitotenv.2020.136951

D'Angelo C, Wiedenmann J (2014) Impacts of nutrient enrichment on coral reefs: new perspectives and implications for coastal management and reef survival. Curr Opin Environ Sustain 7:82–93. https://doi.org/10.1016/j.cosust.2013.11.029

D'Croz L, Mate JL, Oke JE (2001) Responses to elevated seawater temperature and UV radiation in the coral *Porites lobata* from upwelling and non-upwelling environments on the Pacific coast of Panama. Bull Mar Sci 69:203–214

David CP (2003) Heavy metal concentrations in growth bands of corals: a record of mine tailings input through time (Marinduque Island, Philippines). Mar Pollut Bull 46:187–196

De-bashan LE, Bashan Y (2004) Recent advances in removing phosphorous from wastewater and its future use as fertiliser (1997–2003). Water Res 38:4222–4246

DeBose JL, Lema SC, Nevitt GA (2008) Dimethylsulfoniopropionate as a foraging cue for reef fishes. Science 319:1356

De Orte MR, Lombardi AT, Sarmiento AM, Basallote MD, Rodriguez-Romero A, Riba I, delValls A (2014a) Metal mobility and toxicity to microalgae associated with acidification of sediments: CO_2 and acid comparison. Mar Environ Res 96:136–144

De Orte MR, Sarmiento AM, Basallote MD, Rodríguez-Romero A, Riba I, delValls A (2014b) Effects on the mobility of metals from acidification caused by possible CO_2 leakage from sub-seabed geological formations. Sci Total Environ 470–471:356–363. https://doi.org/10.1016/j.scitotenv.2013.09.095

Deschaseaux E, Jones G, Deseo M, Shepherd K, Harrison P, Swan H, Eyre B (2014) Effects of environmental factors on dimethylated Sulphur compounds and their potential role in the antioxidant system of the coral holobiont. Limnol Oceanogr 59(3):758–768

Deschaseaux E, Hardefeldt J, Jones G, Reichelt-Brushett A (2018) High zinc exposure leads to reduced dimethylsulfoniopropionate (DMSP) levels in both the host and endosymbionts of the reef-building coral *Acropora aspera*. Mar Pollut Bull 126:93–100. https://doi.org/10.1016/j.marpolbul.2017.10.070

Diaz RJ, Rosenberg R (2008) Spreading dead zones and consequences for marine ecosystems. Science 321:926–929

Douglas AE (2003) Coral Bleaching-how and why? Mar Pollut Bull 46:385–392

Dove S, Hoegh-Guldberg O, Ranganathan S (2001) Major colour patterns of reef-building corals are due to a family of GFP-like proteins. Coral Reefs 19:197–204. https://doi.org/10.1007/PL00006956

Duckworth A, Giofre N, Jones R (2017) Coral morphology and sedimentation. Mar Pollut Bull 125(1–2):289–300. https://doi.org/10.1016/j.marpolbul.2017.08.036

Dunne RP, Brown BE (2001) The influence of solar radiation on bleaching of shallow water reef corals in the Andaman Sea, 1993–1998. Coral Reefs 20:201–210

El-Naggar HA (2020) Human impacts on coral reef ecosystem. In: Rhodes ER (ed) Natural resources management and biological sciences. https://dx.doi.org/10.5772/intechopen.88841

Erftemeijer PLA, Bernhard R, Hoeksema BW, Todd PA (2012) Environmental impacts of dredging and other sediment disturbances on corals: A review. Mar Pollut Bull 64:1737–1765

Esslemont G (1999) Heavy metals in corals from Heron Island and Darwin Harbour, Australia. Mar Pollut Bull 38:1051–1054

Fabricius KE (2005) Effects of terrestrial runoff on the ecology of corals and coral reefs: review and synthesis. Mar Pollut Bull 50:125–146

Fadlallah YH, Allen KW, Estudillo RA (1995) Mortality of shallow reef corals in the western Arabian Gulf following aerial exposure in winter. Coral Reefs 14(2):99–107

Falkowski PG, Dubinsky Z (1981) Light-shade adaptation of *Stylophora pistillata*, a hermatypic coral from the Gulf of Eilat. Nature 289:172–174. https://doi.org/10.1038/289172a0

Falkowski PG, Dubinsky Z, Muscatine L, Porter JW (1984) Light and the Bioenergetics of a Symbiotic Coral. Bioscience 34:705–709

Falkowski PG, Dubinsky Z, Muscatine L, Mccloskey L (1993) Population control in symbiotic corals. Bioscience 43:606–611

Fallon SJ, White JC, McCulloch MT (2002) Porites corals as recorders of mining and environmental impacts: Misima Island, Papua New Guinea. Geochim Cosmochim Acta 66:45–62

Ferrario F, Beck MW, Storlazzi CD, Micheli F, Shepard CC, Airoldi L (2014) The effectiveness of coral reefs for coastal hazard risk reduction and adaptation. Nat Commun 5:3794. https://doi.org/10.1038/ncomms4794

Fey P, Bustamante P, Bosserelle P, Espiau B, Malau A, Mercader M, Wafo E, Letourneur Y (2019) Does trophic level drive organic and metallic contamination in coral reef organisms? Sci Total Environ 667:208–221. https://doi.org/10.1016/j.scitotenv.2019.02.311

Fine M, Loya Y (2002) Endolithic algae: an alternative source of photo assimilates during coral bleaching. Proc R Soc B Biol Sci 269:1497. https://doi.org/10.1098/rspb.2002.1983

Fisher R, O'Leary RA, Low-Choy S, Mengersen K, Knowlton N, Brainard RE, Caley MJ (2015) Species richness on coral reefs and the pursuit of convergent global estimates. Curr Biol 25:500–505

Fitt WK, Warner ME (1995) Bleaching patterns of four species of Caribbean reef corals. Biol Bull 189:298–307

Fitt WK, Brown BE, Warner ME, Dunne RP (2001) Coral bleaching: interpretation of thermal tolerance limits and thermal thresholds in tropical corals. Coral Reefs 20:51–65

Fonseca JS, Marangoni LFB, Marques JA, Bianchini A (2017) Effects of increasing temperature alone and combined with copper exposure on biochemical and physiological parameters in the zooxanthellate scleractinian coral *Mussismilia harttii*. Aquat Toxicol 190:121–132

Fonseca JS, Marangoni LFB, Marques JA, Bianchini A (2019) Carbonic anhydrase activity as a potential biomarker for acute exposure to copper in corals. Chemosphere 227:598–605. https://doi.org/10.1016/j.chemosphere.2019.04.089

Ganot P, Moya A, Magnone V, Allemand D, Furla P, Sabourault C (2011) Adaptations to endosymbiosis in a cnidarian dinoflagellate association: differential gene expression and specific gene duplications. PLoS Genet 7(7):E1002187. https://doi.org/10.1371/journal.pgen.1002187

Gardener TA, Côté IM, Gill JA, Grant A, Watkinson AR (2003) Long term region-wide declines in Carribbean corals. Science 301:958–960. https://doi.org/10.1126/science.1086050

Garren M, Son K, Raina JB, Rusconi R, Menolascina F, Shapiro OH, Tout J, Bourne DG, Seymour JR, Stocker R (2013) A bacterial pathogen uses dimethylsulphoniopropionate as a cue to target heat-stressed corals. Int Soc Microb Ecol: 1–9

Genin A, Jaffe JS, Reef R, Richter C, Franks PJ (2005) Swimming against the flow: a mechanism of zooplankton aggregation. Science 308:860–862

Glynn PW (1996) Coral reef bleaching: facts, hypotheses and implications. Glob Change Biol 2:495–509

Goreau TJ (1977) Coral skeletal chemistry: physiological and environmental regulation of stable isotopes and trace metals in *Montastrea annularis*. Proc R Soc Lond B 196:291–315

Goreau TF, Goreau NI, Yonge CM (1971) Reef corals: autotrophs or heterotrophs? Biol Bull 141:247–260. https://doi.org/10.2307/1540115

Grottoli AG, Rodrigues LJ, Palardy JE (2006) Heterotrophic plasticity and resilience in bleached corals. Nature 440:1186–1189

Grottoli AG, Warner ME, Levas SJ, Aschaffenburg MD, Schoepf V, McGinley M, Maumann J, Matsui Y (2014) The cumulative impact of annual coral bleaching can turn some coral species winners into losers. Glob Change Biol 20:3823–3833

Guzman HM, Jimenez CE (1992) Contamination of coral reefs by heavy metals along the Caribbean coast of Central-America (Costa Rica and Panama). Mar Pollut Bull 24:554–561

Hamner W, Jones M, Carleton J, Hauri I, Williams DM (1988) Zooplankton, planktivorous fish, and water currents on a windward reef face: Great Barrier Reef, Australia. Bull Mar Sci 42:459–479

Haynes D, Muller J, Carters S (2000) Pesticide and herbicide residues in sediments and seagrasses from the Great Barrier Reef World Heritage Area and Queensland coast. Mar Pollut Bull 41:279–287

Hédouin L, Metain M, Gates RD (2011) Ecotoxicological approach for assessing the contamination of a Hawaiin coral reef ecosystem (Honulua Bay, Maui) by metals and a metalloid. Marine Environmental Research 71:149–161

Hoegh-Guldberg O (1999) Climate change, coral bleaching and the future of the world's coral reefs. Mar Freshw Res 50(8):839–866

Hoegh-Guldberg O (2011) The impact of climate change on coral reef ecosystems. In Coral reefs: an ecosystem in transition by Z Dubinsky and N Stambler (eds). https://doi.org/10.1007/978-94-007-0114-4_22

Hoegh-Guldberg O, Smith GJ (1989) The effect of sudden changes in temperature, light and salinity on the population density and export of zooxanthellae from the reef corals *Stylophora pistillata Esper* and *Seriatopora hystrix Dana*. J Exp Mar Biol Ecol 129(3):279–303. https://doi.org/10.1016/0022-0981(89)90109-3

Hoegh-Guldberg O, Mumby PJ, Hooten AJ, Steneck RS, Greenfield P, Gomez E, Harvell CD, Sale PF, Edwards AJ, Caldeira K, Knowlton N, Eakin CM, Iglesias-Prieto R, Muthiga N, Bradbury RH, Dubi A, Hatziolos ME (2007) Coral reefs under rapid climate change and ocean acidification. Science 318(5857):1737–1742. https://doi.org/10.1126/science.1152509

Hoegh-Guldberg O, Andréfouët S, Fabricius KE, Diaz-Pulido G, Lough JM, Marshall PA, Pratchett MS (2011) Vulnerability of coral reefs in the tropical Pacific to climate change. In: JD Bell, JE Johnson and AJ Hobday (eds) Vulnerability of tropical fisheries and aquaculture to climate change. secretariat of the Pacific Community, Noumea, New Caledonia, Chap 5

Hoegh-Guldberg O, Poloczanska ES, Skiirving W, Dove S (2017) Coral reef ecosystems under climate change and ocean acidification. Front Mar Sci. https://doi.org/10.3389/fmars.2017.00158

Hoegh-Guldberg O, Pendleton L, Kaup A (2019) People and the changing nature of coral reefs. Reg Stud Mar Sci 30:100699. Https://doi.org/10.1016/j.rsma.2019.100699

Holden HM (1999) An analysis of in-situ observations of spectral reflectance characteristics of coral reef features in Fiji and Indonesia. PhD thesis, Department of Geography, University of Waterloo, Canada

Howard LS, Brown BE (1984) Heavy metals and reef corals. Oceanogr Mar Biol Annu Rev 22:195–210

Houlbréque F, Ferrier-Pagés C (2009) Heterotrophy in tropical scleractinian corals. Biol Rev 84:1–17

Hughes TP, Baird AH, Bellwood DR, Card M, Connolly SR, Folke C, Grosberg R, Hoegh-Guldberg O, Jackson JBC, Kleypas J, Lough JM, Marshall P, Nyström M, Palumbi SR, Pandolfi JM, Rosen B, Roughgarden J (2003) Climate change, human impacts, and the resilience of coral reefs. Science 301:929–933. https://doi.org/10.1126/science.1085046

Hughes AD, Grottoli AG, Pease TK, Matsui Y (2010) Acquisition and assimilation of carbon in non-bleached and bleached corals. Mar Ecol-Prog Series 420:91–101. https://doi.org/10.3354/meps08866

Hughes AD, Grottoli AG (2013) Heterotrophic compensation: a possible mechanism for resilience of coral reefs to global warming or a sign of prolonged stress? PLoS ONE 8(11): https://doi.org/10.1371/journal.pone.0081172

Hughes TP, Kerry JT, Alvarez-Noriega M, Alvarez-Romero JG, Anderson KD, Baird AH et al (2017) Global warming and recurrent mass bleaching of corals. Nature 543:373–377. https://doi.org/10.1038/nature21707

Hughes TP, Anderson KD, Connolly SR, Heron SF, Kerry JT, Lough JM et al (2018) Spatial and temporal patterns of mass bleaching of corals in the Anthropocene. Science 359:80

Hume BCC, D'Angelo C, Smith EG, Stevens JR, Burt J, Wiedenmann J (2015) Symbiodinium thermophilum sp. nov., a thermotolerant symbiotic alga prevalent in corals of the world's hottest sea, the Persian/Arabian Gulf. Sci Rep 5:8562. https://doi.org/10.1038/srep08562

IPCC (2007) https://www.ipcc.ch/pdf/assessment-report/ar4/syr/ar4_syr_full_report.pdf

IPCC (2013) "Climate change 2013: the physical science basis. In: Stocker TF, Qin D, Plattner G-K, Tignor M, Allen SK, Boschung J et al (eds) Contribution of Working Group I to the Fifth Assessment Report of the Intergovernmental Panel on climate change. (Cambridge; New York, NY: Cambridge University Press)

Ishimatsu A, Hayashi M, Kikkawa T (2008) Fishes in high-CO_2 acidified oceans. Mar Ecol Prog Ser 373:295–302

Ivanina AV, Sokolova IM (2015) Interactive effects of metal pollution and ocean acidification on physiology of marine organisms. Curr Zool 61(4):653–668

Johnson JE, Marshall PA (2007) Climate change and the great barrier reef. A vulnerability assessment. Great Barrier Reef Marine Park Authority and Australian Greenhouse Office, Australia

Jokiel PL, Coles SL (1990) Response of Hawaiian and other Indo-Pacific reef corals to elevated temperature. Coral Reefs 8:155–162

Jones RJ (1997) Zooxanthallae loss as a bioassay for assessing stress in corals. Mar Ecol Prog Ser 149:163–171

Jones GB, King S (2015) Dimethylsulphoniopropionate (DMSP) as an Indicator of bleaching tolerance in Scleractinian Corals. J Mar Sci Eng 3:444–465. https://doi.org/10.3390/jmse3020444

Jones RJ, Kildea T, Hoegh-Guldberg O (1999) PAM chlorophyll fluorometry: a new insitu technique for stress assessment in scleractinian corals, used to examine the effects of cyanide from cyanide fishing. Mar Pollut Bull 38:864–874

Jones GB, Curran MAJ, Broadbent AD, King S, Fischer E, Jones R (2007) Factors affecting the cycling of dimethylsulphide and dimethylsulphoniopropionate in coral reef waters of the Great Barrier Reef. Environ Chem 4:310–322

Jones GB, Fischer E, Deschaseaux ESM, Harrison PL (2014) The effect of coral bleaching on the cellular concentration of dimethylsulphoniopropionate in reef corals. J Exp Mar Biol Ecol 460:19–31. https://doi.org/10.1016/j.jembe.2014.06.003

Jones R, Ricardo GF, Negri AP (2015) Effects of sediments on the reproductive cycle of corals. Mar Pollut Bull 100(1):13–33. https://doi.org/10.1016/j.marpolbul.2015.08.021

Jupiter S, Roff G, Marion G, Henderson M, Schrameyer V, McCulloch M, Hoegh-Guldberg O (2008) Linkages between coral assemblages and coral proxies of terrestrial exposure along a cross-shelf gradient on the southern Great Barrier Reef. Coral Reefs 27:887–903. https://doi.org/10.1007/s00338-008-0422-3

Kirst GO, Thiel C, Wolff H, Nothnagel J, Wanzek M, Ulmke R (1991) Dimethylsulfoniopropionate (DMSP) in ice-algae and its possible biological role. Mar Chem 35:381–388

Knight K (2012) Hatchling loggerhead turtles pick up DMS. J Exp Biol 215(III). https://doi.org/10.1242/jeb.080044

Lana A, Bell TG, Simó R, Vallina SM, Ballabrera-Poy J, Kettle AJ, Liss PS (2011) An updated climatology of surface dimethylsulfide concentrations and emission fluxes in the global ocean. Global Biogeochem Cycles 25(1):1–17. https://doi.org/10.1029/2010GB003850

Langdon C, Atkinson MJ (2005) Effect of elevated pCO2 on photosynthesis and calcification of corals and interactions with seasonal change in temperature/irradiance and nutrient—enrichment. J Geophys Res Oceans 110:C9. https://doi.org/10.1029/2004JC002576

Lebrec M, Stefanski S, Gates R, Acar S, Golbuu Y, Claudel-Rusin A, Kurihara H, Rehdanz K, Paugam-Baudoin D, Tsunoda T, Swarzenski PW (2019) Ocean acidification impacts in select Pacific Basin coral reef ecosystems. Reg Stud Mar Sci 28: https://doi.org/10.1016/j.rsma.2019.100584

Lesser MP (1997) Oxidative stress causes coral bleaching during exposure to elevated temperatures. Coral Reefs 16:187–192

Lesser MP (2004) Experimental Biology of coral reef ecosystems. J Exp Mar Biol Ecol 300:217–252

Levas S, Grottoli AG, Schoepf V, Aschaffenburg M, Baumann J, Bauer JE, Warner ME (2015) Can heterotrophic uptake of dissolved organic carbon and zooplankton mitigate carbon budget deficits in annually bleached corals? Coral Reefs 35(2):495–506. https://doi.org/10.1007/s00338-015-1390-z

Lewis SE, Brodie JE, Bainbridge ZT, Rohde KW, Davis AM, Masters BL, Maughan M, Devlin MJ, Mueller JF, Schaffelke B (2009) Herbicides: a new threat to the Great Barrier Reef. Environ Pollut 157:2470–2484

Li J, Sun C, Zhang L, Ding J, Jiang F, Wang Z, Wang Z, Fu L (2020) Current distribution characteristics of trace elements in the coral-reef systems of Xisha Islands, China. Mar Pollut Bull 150: https://doi.org/10.1016/j.marpolbul.2019.110737

Luschak VI (2011) Environmentally induced oxidative stress in aquatic animals. Aquat Toxicol 101:13–30

Madin JS, Hoogenboom MO, Connolly SR, Darling ES, Falster DS, Huang D, Keith SA, Mizerek T, Pandolfi JM, Putnam HM, Baird AH (2016) A trait-based approach to advance coral reef science. Trends Ecol Evol 31(6). http://dx.doi.org/10.1016/j.tree.2016.02.012

Marangoni LFB, Pinto MMdA, Marques JA, Bianchini A (2019) Copper exposure and seawater acidification interaction: Antagonistic effects on biomarkers in the zooxanthellate scleractinian coral Mussismilia harttii. Aquat Toxicol 206:123–133. https://doi.org/10.1016/j.aquatox.2018.11.005

Marques JA, Abrantes DP, Marangoni LFB, Bianchini A (2019) Ecotoxicological responses of a reef calcifier exposed to copper, acidification and warming: A multiple biomarker approach. Environ Pollut. https://doi.org/10.1016/j.envpol.2019.113572

Maximillian J, Brusseau ML, Glenn EP, Matthias AD (2019) Pollution and environmental perturbations in the global system. In: Brusseau ML, Pepper IL, Gerba CP (eds) Environmental and pollution science, Chap 25, 3rd edn, pp 457–476. https://doi.org/10.1016/B978-0-12-814719-1.00025-2

McCulloch M, Fallon S, Wyndham T, Hendy E, Lough J, Barnes D (2003) Coral record of increased sediment flux to the inner Great Barrier Reef since European settlement. Nature 421(6924):727–730

McGregor HV, Fischer MJ, Gagan MK, Fink D, Phipps SJ, Wong H, Woodroffe CD (2013) A weak El Nino/Southern Oscillation ~ with delayed seasonal growth around 4,300 years ago. Nat Geosci 6:949–953. https://doi.org/10.1038/ngeo1936

Mieog JC, Olsen JL, Berkelmans R, Bleuler-Martinez SA, Willis BL, van Oppen MJH (2009) The Roles and interactions of symbiont, host and environment in defining coral fitness. PLoS ONE 4(7): https://doi.org/10.1371/journal.pone.0006364

Millero FJ, Woosley R, DiTrolio BR, Waters J (2009) Effects of the ocean acidification on the speciation of metals in seawater. Oceanography 22:72–85

Mollica NR, Guo W, Cohen AL, Huang KF, Foster GL, Donald HK, Solow AR (2018) Ocean acidification affects coral growth by reducing skeletal density. Proc Natl Acad Sci U SA 115(8):1754–1759. www.pnas.org/cgi/doi/10.1073/pnas.1712806115

Monserrat JM, Martínez PE, Geracitano LA, Amado LL, Martins CM, Pinho GL, Chaves IS, Ferreira-Cravo M, Ventura-Lima J, Bianchini A (2007) Pollution biomarkers in estuarine animals: critical review and perspectives. Comp Biochem Physiol C 146:221–234

Muller-Parker G, D'Elia CF, Cook CB (2015) Interactions between corals and their symbiotic algae. In Birkeland C (Ed) Coral reefs in the anthropocene. Springer, Dordrecht. https://doi.org/10.1007/978-94-017-7249-5_5

Muscatine L (1990) The role of symbiotic algae in carbon and energy flux in reef corals. In: Dubinsky Z (ed) Ecosystems of the world: coral reefs. Elsevier, Amsterdam, pp 75–87

Muscatine L, Porter JW (1977) Reef corals: mutualistic symbioses adapted to nutrient poor environments. Bioscience 27:454–460

Muscatine L, Mccloskey LR, Marian RE (1981) Estimating the daily contribution of carbon from zooxanthellae to coral animal respiration. Limnol Oceanogr 26:601–611

Muscatine L, Falkowski PG, Dubinsky Z (1983) Carbon budgets in symbiotic associations. In: Schenk HEA, Schwemmler W (eds) Endocytobiology II, intracellular space as oligogenetic ecosystem. Walter DeGruyter, Berlin, pp 649–658

Negri AP, Hoogenboom MO (2011) Water contamination reduces the tolerance of coral larvae to thermal stress. PLoS ONE 6: https://doi.org/10.1371/journal.pone.0019703

Negri AP, Harford AJ, Parry DL, van Dam RA (2011) Effects of alumina refinery wastewater and signature metal constituents at the upper thermal tolerance of: 2. The early life stages of the coral Acropora tenius. Mar Pollut Bull 62:474–482

Nishiguchi MK, Somero GN (1992) Temperature and concentration dependence of compatibility of the organic osmolyte B-dimethylsulfoniopropionate. Cryobiology 129:118–124

Orr JC, Fabry VJ, Aumont O, Bopp L, Doney SC, Feely RA, Gnanade-sikan A, Gruber N, Ishida A, Joos F, Key RM, Lindsay K, Maier-Reimer E, Mateat R, Monfray P, Mouchet A, Najjar RG, Plattner GK, Rodgers KB, Sabine CL, Sarmiento JL, Schlitzer R, Slater RD, Totter-dell IJ, Weirig MF, Yamanaka Y, Yool A (2005) Anthropogenic ocean acidification over the twenty-first century and its impact on calcifying organisms. Nature 437(681–686)

Otte ML, Morris JT (1994) Dimethylsulphoniopropionate (DMSP) in *Spartina alterniflora Loisel.* Aquat Bot 48:239–259

Palardy JE, Rodrigues LJ, Grottoli AG (2008) The importance of zooplankton to the daily metabolic carbon requirements of healthy and bleached corals at two depths. J Exp Mar Biol Ecol 367:180–188

Pandolfi JM, Bradbury RH, Sala E, Hughes TP, Bjorndal KA, Cooke RG, McArdle D, McClenachan L, Newman MJH, Paredes G, Warner RR, Jackson JBC (2003) Global trajectories of the long term decline of coral reef ecosystems. Science 301:955–958

Perez SF, Cook CB, Brooks WR (2001) The role of symbiotic dinoflagellates in the temperature-induced bleaching response of the subtropical sea anemone *Aiptasia pallida.* J Exp Mar Biol Ecol 256:1–14

Perry CT, Kench PS, O'Leary MJ, Morgan KM, Januchowski-Hartley F (2015) Linking reef ecology to island building: parrotfish identified as major producers of island building sediment in the Maldives. Geology 43(6):503–506. https://doi.org/10.1130/G36623.1

Peters EC, Gassman NJ, Firman JC, Richmond RH, Power EA (1997) Ecotoxicology of tropical marine ecosystems. Environ Toxicol Chem 16:12–40

Pogoreutz C, Radecker N, Cardenas A, Gardes A, Voolstra CR, Wild C (2017) Sugar enrichment provides evidence for a role of nitrogen fixation in coral bleaching. Glob Change Biol 23:3838–3848. https://doi.org/10.1111/gcb.13695

Porter JW (1976) Autotrophy, heterotrophy, and resource partitioning in Caribbean reef-building corals. Am Nat 110(975):731–742

Pörtner HO, Langenbuch M, Michaelidis B (2005) Synergistic effects of temperature extremes, hypoxia, and increases in CO_2 on marine animals: from Earth history to global change. J Geophys Res 110:C9. https://doi.org/10.1029/2004JC002561

Prouty NG, Hughen KA, Carilli J (2008) Geochemical signature of land-based activities in Caribbean coral surface samples. Coral Reefs 27:727–742

Prouty NG, Field ME, Stock JD, Jupiter SD, McCulloch M (2010) Coral Ba/Ca records of sediment input to the fringing reef of the Southshore of Moloka'i, Hawai'i over the last several decades. Mar Pollut Bull 60:1822–1835. https://doi.org/10.1016/j.marpolbul.2010.05.024

Putnam HM, Barott KL, Ainsworth TD, Gates RD (2017) The vulnerability and resilience of reef-building corals. Curr Biol 27:11. https://doi.org/10.1016/j.cub.2017.04.047

Raina J-B (2013) Production and fate of dimethylsulfoniopropionate (DMSP) in reef-building corals and its integral role in coral health. PhD thesis, James Cook University

Raina J-B, Tapiolas DM, Forêt S, Lutz A, Abrego D, Ceh J, Seneca FO, Clode PL, Bourne DG, Willis BL, Motti CA (2013) DMSP biosynthesis by an animal and its role in coral thermal stress response. Nature. https://doi.org/10.1038/nature12677

Reaka-Kudla ML (1997) Global biodiversity of coral reefs: a comparison with rainforests. In: Reaka-Kudla ML, Wilson DE (eds) Biodiversity II: understanding and protecting our biological resources. Joseph Henry Press. Reaka-Kudla ML 2001. http://www.nap.edu/openbook/030905 2270/html85.html

Reaka-Kudla ML (2001) Known and unknown biodiversity, risk of extinction, and conservation strategy in the sea. In: Bendell-Young L, Gallaugher P (eds) Waters in peril. Kluwer Academic Publishers, NY, pp 19–33

Reynaud S, Ferrier-Pagès C, Sambrotto R, Juillet-Leclerc A, Jaubert J et al (2002) Effect of feeding on the carbon and oxygen isotopic composition in the tissues and skeleton of the zooxanthellate coral *Stylophora pistillata.* Mar Ecol Prog Ser 238:81–89. https://doi.org/10.3354/meps238081

Ricardo GF, Jones RJ, Clode PL, Humanes A, Giofre N, Negri AP (2018) Sediment characteristics influence the fertilization success of the corals *Acropora tenuis* and *Acropora millepora*. Mar Pollut Bull 135:941–953

Rodrigues LJ, Grottoli AG (2007) Energy reserves and metabolism as indicators of coral recovery from bleaching. Limnology and Oceanograpy 52(5):1874–1882

Rogers CS (1990) Responses of coral reefs and reef organisms to sedimentation. Mar Ecol Prog Ser 62:185–202

Rogers CS (1993) Hurricanes and Coral Reefs: The intermediate disturbance hypothesis revisited. Coral Reefs 12:127–137. https://doi.org/10.1007/BF00334471

Ruiz-Torres V, Encinar JA, Herranz-Lopez M, Perez-Sanchez A, Galiano V, Barrajon-Catalan E, Micol V (2017) An updated review on marine anticancer compounds: the use of virtual screening for the discovery of small-molecule cancer drugs. Molecules 22:1037

Runnalls LA, Coleman ML (2003) Record of natural and anthropogenic changes in reef environments (Barbados West Indies) using laser ablation ICP-MS and sclerochronology on coral cores. Coral Reefs 22:416–426

Savoca MS, Nevitt GA (2014) Evidence that dimethyl sulphide facilitates a tritrophic mutualism between marine primary producers and top predators. In: Proceedings of national academy of sciences U. S. A. (in press)

Schwarz JA, Mitchelmore CL, Jones R, O'Dea A, Seymour S (2013) Exposure to copper induces oxidative and stress responses and DNA damage in the coral *Montastraea franksi*. Comp Biochem Physiol Part C Toxicol Pharmacol 157(3):272–279. https://doi.org/10.1016/j.cbpc.2012.12.003

Seeman J (2013) The use of ^{13}C and ^{15}N isotope labeling techniques to assess heterotrophy of corals. J Exp Mar Biol Ecol 442:88–95

Seymour JR, Simo R, Ahmed T, Stocker R (2010) Chemoattraction to dimethylsulfoniopropionate throughout the marine microbial food web. Science: 342–345

Shah S (2008) Study of heavy metal accumulation in *scleractinian* corals of Viti Levu, Fiji Islands, unpublished. MSc thesis, Faculty of Science, Technology and Environment, The University of the South Pacific, Fiji

Shick JM, Lesser MP, Jokiel PJ (1996) Effects of ultraviolet radiation on corals and other coral reef organisms. Glob Change Biol 2:527–545

Siddiqui S, Bielmyer GK (2015) Responses of the sea anemone, *Exaiptasia pallida*, to ocean acidification conditions and copper exposure. Aquat Toxicol 167:228–239. https://doi.org/10.1016/j.aquatox.2015.08.012

Sieburth JM (1960) Acrylic acid an "antibiotic" principle in Phaeocystis blooms in Antarctic waters. Science 132:676–677

Sieburth JM (1961) Antibiotic properties of acrylic acid a factor in the gastro-intestinal antibiosis of polar marine animals. J Bacteriol 82:72–79

Smith DJ, Suggett DJ, Baker NR (2005) Is photoinhibition of zooxanthellae photosynthesis the primary cause of thermal bleaching in corals? Glob Change Biol 11:1–11

Sokolova IM, Matoo O, Dickinson G, Beniash E (2015) Physiological effects of ocean acidification on animal calcifiers. In: Whiteley SN, Solan M (eds) Stressors in the marine environments. Oxford University Press, Oxford

Souter DW and Lindén O (2000) The health and future of coral reef systems. Ocean and Coastal Management, 43: 657–688

Spalding MD, Ravilious C, Green EP, UNEP-WCMC (2001) World Atlas of coral reefs

Spalding M, Burke L, Wood SA, Ashpole J, Hutchisone J, zu Ermgassene P (2017) Mapping the global value and distribution of coral reef tourism. Marine Policy 82:104–113. http://dx.doi.org/10.1016/j.marpol.2017.05.014

Stefels J (2000) Physiological aspects of the production and conversion of DMSP in marine algae and higher plants. J Sea Res 43:183–197

Storlazzi CD, Norris BK, Rosenberger KJ (2015) The influence of grain size, grain color, and suspended-sediment concentration on light attenuation: why fine grained terrestrial sediment is bad for coral reef ecosystems. Coral Reefs 34(3):967–975. https://doi.org/10.1007/s00338-015-1268-0

Sunda W, Keiber DJ, Kiene RP, Huntsman S (2002) An antioxidant function for DMSP and DMS in marine algae. Nature 418:317–319

Swan H, Jones G, Deschaseux E (2012) Dimethylsulfide and coral reef ecosystems. In: 12th international coral reef symposium, Cairns, 9–13 July

Tagliafico A, Rangel S, Kelaher B, Christidis L (2018) Optimizing heterotrophic feeding rates of three commercially important scleractinian corals. Aquaculture 483:96–101

Teh LSL, Teh LCL, Sumaila UR (2013) A global estimate of the number of coral reef fishers. PLoS ONE 8: https://doi.org/10.1371/journal.pone.0065397

Tremblay P, Grover R, Maguer JF, Legendre L, Ferrier-Pagés C (2012) Autotrophic carbon budget in coral tissue: a new ^{13}C-based model of photosynthate translocation. The Journal of Experimental Biology 215:1384–1393

The Royal Society (2005) Ocean acidification due to increasing atmospheric carbon dioxide. http://www.us-ocb.org/publications/Royal_Soc_OA.pdf

UNEP (2005) Coral Reef Management. http://www.cep.unep.org/marine-issues/plonearticlemultip lepage.2005

UNEP/IUCN (1988) Coral reefs of the world. vol. 2: Indian Ocean, Red Sea and Gulf. UNEP Regional Seas Directories and Bibliographies. IUCN, Gland, Switzerland and Cambridge, UK. UNEP, Nairobi, Kenya, 389 pp

U.S. Environmental Protection Agency (2016) Climate change indicators in the United States, 2016, 4th edn. EPA 430-R-16-004. www.epa.gov/climate-indicators

Vairavamurthy A, Andreae MO, Iverson RL (1985) Biosynthesis of dimethylsulfide and dimethyl-propiothetin by *Hymenomonas carterae* in relation to sulfur source and salinity variations. Limnol Oceanogr 30:59–70

Van Alstyne KL, Wolfe GV, Freidenburg TL, Neill A, Hicken C (2001) Activated defense systems in marine macroalgae: evidence for an ecological role for DMSP cleavage. Mar Ecol Prog Ser 213:53–65

Van Alstyne KL, Houser LT (2003) Dimethylsulfide release during macroinvertebrate grazing and its role as an activated chemical defense. Mar Ecol Prog Ser 250:175–181

Van Alstyne KL, Dominique VJI, Muller-parker G (2008) Is dimethylsulfoniopropionate (DMSP) produced by the symbionts or the host in an anemone-zooxanthella symbiosis? Coral Reefs 28:167–176

Venn AA, Loram JE, Douglas AE (2008) Photosynthetic symbiosis in animals. J Exp Bot 59(5):1069–1080. https://doi.org/10.1093/jxb/erm328

Veron JEN (1995) Corals in space and time: the biogeography and evolution of the *Scleractinia*. Cornell University Press, Ithaca, NY

Veron JEN (1986) Corals of Australia and Indo-Pacific. Angus and Robertson, London, 644 pp

Wang Z, Wang Y, Zhao P, Chen L, Yan C, Yan Y, Chi Q (2015) Metal release from contaminated coastal sediments under changing ph conditions: Implications for metal mobilization in acidified oceans. Mar Pollut Bull 101:707–715. https://doi.org/10.1016/j.marpolbul.2015.10.026

Wei G, McCulloch MT, Mortimer G, Deng W, Xie L (2009) Evidence for ocean acidification in the Great Barrier Reef of Australia. Geochim Cosmochim Acta 73:2332–2346. https://doi.org/10.1016/j.gca.2009.02.009

Wei WC, Sung PY, Duh CY, Chen BW, Sheu JH, Yang NS (2013) Anti-inflammatory activities of natural products isolated from soft corals of Taiwan between 2008 and 2012. Mar Drugs 11:4083–4126

Weijerman M, Veazey L, Yee S, Vaché K, Delevaux JMS, Donovan MK, Falinski K, Lecky J, Oleson KLL (2018) Managing local stressors for coral reef condition and ecosystem services delivery under climate scenarios. Frontiers in Marine Science 5:425. https://doi.org/10.3389/fmars.2018.00425

Weis VM (2008) Cellular mechanisms of Cnidarian bleaching: stress causes the collapse of symbiosis. J Exp Biol 211:3059–3066

Wilde KL, Stauber JL, Markich SJ, Franklin NM, Brown PL (2006) The effect of Ph on the uptake and toxicity of copper and zinc in a tropical freshwater alga (Chlorella sp). Arch Environ Contam Toxicol 51:174–185

Wilkinson CR (2004) Status of coral reefs of the world: 2004. Australian Institute of Marine Science, Cape Ferguson, Qld, vol 1, 301 pp

Williams GJ, Sandin SA, Zgliczynski BJ, Fox MD, Gove JM, Rogers JS, Furby KA, Hartmann AC, Caldwell ZR, Price NN, Smith JE (2018) Biophysical drivers of coral trophic depth zonation. Mar Biol 165:60. https://doi.org/10.1007/s00227-018-3314-2

Wolfe GV, Steinke M, Kirst GO (1997) Grazing-activated chemical defence in a unicellular marine alga. Nature 387:894–897

Yang Y, Hansson L, Gattuso JP (2016) Data compilation on the biological response to ocean acidification: an update. Earth Syst Sci Data 8:79–87. 10.5194/essd-8-79-2016. www.earth-syst-sci-data.net/8/79/2016/

Yost DM, Jones RJ, Mitchelmore CL (2010) Alterations in dimethylsulfoniopropionate (DMSP) levels in the coral *Montastrea franski* in response to copper exposure. Aquat Toxicol 98(4):367–373. https://doi.org/10.1016/j.aquatox.2010.03.005

Zeng X, Chen X, Zhuang J (2015) The positive relationship between ocean acidification and pollution. Mar Pollut Bull 91:14–21. https://doi.org/10.1016/j.marpolbul2014.12.001

Zhang XH, Liu J, Liu J, Yang G, Xue CX, Curson ARJ, Todd JD (2019) Biogenic production of DMSP and its degradation to DMS-their roles in the global sulfur cycle. Sci China Life Sci Spec Topic Elem Cycl Microorg Hydrosp 62(10):1296–1319. https://doi.org/10.1007/s11427-018-9524-y

Zinke J, Rountrey A, Feng M, Xie S-P, Dissard D, Rankenburg K, Lough JM, McCulloch MT (2014) Coral record long-term Leeuwin current variability including Ningaloo Nino/Nina since 1795. Nat Commun 5:3607. https://doi.org/10.1038/ncomms4607

Chapter 3
The Use of *Scleractinian* Corals for Heavy Metal Studies

Abstract *Scleractinian* corals belong to the phylum Cnidaria, class Anthozoa and order *Scleractinia* and are the architects of coral reefs; therefore a good representative of environmental changes. Their skeletons assimilate records of the presence and concentrations of certain metals over centuries without being killed by the relatively high levels present in the marine environment. Corals reproduce both sexually and asexually. Environmental factors such as water chemistry, sedimentation, pollution and surface temperature induce *scleractinian* corals to incorporate the heavy metals in their living parts and skeletons. Heavy metals get incorporated into the coral skeleton through substitution of dissolved metal species via calcium substitution and once incorporated into the coral skeleton, the metals remain embedded forever since the new growth covers the old carbonate surface. Once incorporated within the coral, heavy metals can result in acute or chronic toxicity causing lethal effects or long-term impacts to key biological processes of corals.

Keywords Scleractinian corals · Coral reproduction · Calcium substitution · Heavy metals

3.1 *Scleractinian* Corals

Scleractinian corals belong to the phylum Cnidaria; class Anthozoa and order *scleractinia*. *Scleractinian* corals have a mutualistic relationship with endosymbiotic dinoflagellates in the genus *Symbiodinium* (referred to as 'zooxanthallae'-discussed in Chap. 2) and are the architects of coral reefs. Being sedentary in nature, they become good representatives and indicators of their environment and its related environmental changes and thus are able to provide valuable information. Boboria et al. (2021) agree that stationary sedentary organisms are more beneficial to use than the mobile organisms such as fish, when dealing with heavy metals in the marine environment. In addition to this, these corals are ample throughout the study area, are able to accumulate the pollutants without being killed by comparatively high levels encountered in the marine environment and are sufficiently long lived, thus adequate skeleton can be obtained for analysis and thin sections (El-Sorogy et al.

2012). Environmental parameters such as SSTs, precipitation, salinity, sedimentation and pollution can easily be obtained from the coral records.

Peters et al. (1997) mentioned that *scleractinian* corals and particularly those of the genus *Acropora* have been identified as sensitive to changes in water quality and chemical contaminants; making them an appropriate representative tropical test species for use in investigating potential risks from contamination to coral reefs. Scleractinian corals (particularly their skeletons) are used as proxy tools to record environmental conditions/pollution such as trace metals (Guzman and Jimenez 1992; Bastidas and Garcia 1999; Esslemont 1999; Fallon et al. 2002; David 2003; Runnalls and Coleman 2003; Khaled et al. 2003; Ramos et al. 2004). Their skeletons assimilate records of the presence and concentrations of certain metals over centuries, without being killed by the relatively high levels present in the marine environment.

Since the trace metal concentrations in the skeletons reveal the environmental metal loads, hence *scleractinian* corals become an excellent environmental indicator for pollution studies. Prouty et al. (2013) highlighted that since corals precipitate their $CaCO_3$ skeleton directly from seawater, the geochemistry of coral skeletons had been widely used to study environmental variability. At the time of accretion of its $CaCO_3$ skeleton, corals obtain an unaltered record of the chemical and physical conditions that exist in the surrounding seawater and this record becomes useful in monitoring studies.

One of the advantages of corals being an oceanic recorder is the availability of the enhanced time resolution (biweekly to seasonal) records obtainable from the high growth rate and according to Prouty et al. (2013), these high-resolution proxy records provide detailed information on environmental change by accurately recording variations in physical and chemical parameters such as SST, river input, salinity and pollution. It is important to note that corals are a well-established indicator taxon of metal pollution (Schyff et al. 2020); and that metal pollution studies are best conducted on sessile or sedentary organisms to ensure some certainty about exposure rate (Reichelt-Brushett 2012). Coral reefs are gradually exposed to a variety of anthropogenic stressors (Gissi et al. 2019) and metals may contribute to the decline in the health of coral reef ecosystems (Mitchelmore et al. 2007). The metal contaminants originate from a range of sources including industrial effluent, agricultural runoff and antifouling paints (van Dam et al. 2011). Rainbow (1995) highlighted that marine pollution increases the concentration of natural metals beyond background concentration to levels that may harm organisms.

3.2   Reproduction in *Scleractinian* Corals

Corals are the principal structural architects of coral reef ecosystems and therefore coral reproduction becomes an essential process that sustains the maintenance of the structure and functions of the coral reef ecosystems (Baird et al. 2015). Coral reproduction is controlled by several life processes such as gamete production, fertilization, planktonic larval dispersal, larval settlement, post-settlement growth, and survival (Ceh et al. 2012).

Corals depict a simple life cycle involving two phases: a dominant benthic polyp phase and a shorter planula larval phase. The polyp phase is characterized by growth of tissues and skeleton that often includes one or more forms of asexual budding or reproduction; and repeated cycles of sexual reproduction involving the production of gametes, fertilization, embryo development, and a larval phase that is usually planktonic and dispersive to some degree (Harrison and Wallace 1990). Survival of the planula and successful attachment on a hard substratum enables it to form into a juvenile polyp which initiates the formation of the calcium carbonate exoskeleton. Subsequent growth during an initial presexual juvenile phase leads to development of the adult form that becomes sexually reproductive, which completes the life cycle (Harrison and Wallace 1990).

Majority of the reef-building corals reproduce sexually and asexually (Combosch and Vollmer 2013). Sexual reproduction allows genetic recombination and production of new coral genotypes, thus enhancing fitness and survival of the species (Harrison 2011). Sexual reproduction occurs by either free spawning eggs and sperm or internally brooding larvae inside the coral polyp (Harrison 2011) with the resulting planula dispersing, settling and metamorphosing to form new coral polyps; which grow by asexual fission of individual polyps (Combosch and Vollmer 2013).

Asexual reproduction; referred to as fragmentation; occurs when new colonies form from branching coral, which are broken by physical forces such as wave action. Asexual reproduction produces genetically identical modules that may prolong the survival of the genotype (Harrison 2011).

Corals could be hermaphroditic or gonochoric in nature and thus Harrison (2011) reported that four basic patterns of sexual reproduction are evident among corals, which include: hermaphroditic broadcast spawners, hermaphroditic brooders, gonochoric broadcast spawners and gonochoric brooders.

Hermaphroditic corals consist of both sexes developed in their polyps and colonies, whereas gonochoric corals have separate sexes; and corals with these sexual patterns either broadcast spawn their gametes for external fertilization and subsequent embryo and larval development, or have internal fertilization and brood embryos and planula larvae within their polyps (reviewed by Harrison and Wallace 1990). In case of hermaphroditic coral species, eggs and spermaries develop either on the same mesenteries (such as in the faviids and mussids) or on different mesenteries within the same polyp (as in most pocilloporids and acroporids) or in different polyps within the same colony (Fadlallah 1983, Harrison and Wallace 1990). For example, Ward and Harrison (2000) mentioned that Acroporids are hermaphroditic with eggs and spermaries on different mesenteries within the same polyp.

Harrison (2011) and Kerr et al. (2011) highlighted that approximately 70–75% of corals are hermaphroditic, while remaining 25–30% gonochoric. Oocytes and testes develop within the same polyp in hermaphroditic corals whilst in gonochoric corals male colonies produce sperm and female colonies produce oocytes (Hidaka 2016).

Many scleractinian corals reproduce by broadcast spawning and are referred to as broadcast spawners. There are release of gamete bundles from different coral colonies into the water column such that external fertilization, larval development and dispersal take place (Harrison et al. 1984). Harrison (2011) and Kerr et al. (2011) reported that about 87–90% of corals release sperm and /or eggs to seawater and thus fertilisation occurs within the water column.

Mass spawning involves the synchronous release of gametes and is dependent on the geographical location, occurring in the summer, once a year over a few nights following the full moon (Harrison et al. 1984). For instance, throughout annual mass spawning events on the Great Barrier Reef, Australia, up to 150 coral species release their gametes synchronously into the water column (Reichelt-Brushett and Michalek-Wagner 2005). Within the first 24 h, the planula larvae of broadcast spawning corals develop and generally settle onto the substrate within 3–5 days of spawning (Babcock and Heyward 1986). Synchronous spawning among hermatypic corals maximises reproductive success and allows genetic exchange among otherwise asexual colonies (Harrison et al. 1984).

In brooding corals, fertilization occurs within the polyp where the fertilized eggs develop to form the planula larvae within the gastrovascular cavity of the polyp and this usually occurs within the remaining 10–13% of the corals (Hidaka 2016). Nonetheless, in both spawners and brooders, the planula larvae transform to polyps after settlement (Hidaka 2016). Natural or anthropogenic stressors such as increase in water temperature, coral bleaching, or pollution can influence the reproductive cycles and sustainability of coral gametes (Ward et al. 2002; Harrison and Ward 2001; Hudspith et al. 2017).

3.3 Biogeochemistry of *Scleractinian* Corals

Environmental factors such as water chemistry, sedimentation, pollution and surface temperature induce *scleractinian* corals to incorporate the heavy metals in their living parts and skeletons. According to Al-Rousan et al. (2007), heavy metals occur in the coral skeletons as a result of structural incorporation of metals into the aragonite (Goreau 1977), inclusion of particulate materials in skeletal cavities (reviewed by Howard and Brown 1984), surface adsorption onto exposed skeleton (St John 1974; Brown et al. 1991), and chelation with the organic matrix of the skeleton (Mitterer 1978).

Heavy metals are incorporated into the crystal lattice coral skeleton through substitution of dissolved metal species via calcium substitution (Ferrier-Pages et al.2005). Howard and Brown (1984) highlighted that through feeding, corals incorporate heavy metals present in seawater into their tissue and skeleton either as dissolved ion or as particulate forms. The dissolved ions are taken up by the zooxanthallae and tissue and later transferred to the skeleton whilst the particulate forms are ingested by the plankton and transferred to the tissue (Howard and Brown 1984). The uptake of metals from the aquatic environment is usually through the living organism and then translocated to the nonliving component (skeleton); with the incorporation of metals into the skeleton matrix being dependent on the patterns of uptake and regulation in corals (Hardefeldt and Reichelt-Brushett 2015).

Once incorporated into the coral skeleton, the metals will remain embedded forever since the new growth covers the old carbonate surface (St John 1974). Hence, Shen and Boyle (1988) and Esslemont et al. (2000) mention that with continuous changes in the marine environments, most studies have examined the accumulation of heavy metals in coral tissues and skeletons to determine the long-term effects of environmental factors in addition to direct anthropogenic inputs. Scleractinian corals are the dominant structuring organisms of coral reefs, hence it becomes imperative to understand the potential effects of trace metals on their health and reproductive success (Reichelt-Brushett and Harrison 2005).

3.4 Ecotoxicological Studies Using *Scleractinian* Corals

Toxicology comprises the study of the harmful effects of toxins and anthropogenic compounds (such as heavy metals) on living organisms. Rotchell and Ostrander (2011) highlighted that focused toxicological studies at the subcellular/molecular level of corals have the potential to make accessible a great deal of valuable information. Ecotoxicological testing is an operative management tool which is used to evaluate the potential toxicity of environmental contaminants such as heavy metals (Peters et al. 1997; van Dam et al. 2008). For instance, ecotoxicology provides a tool in a multiple lines of evidence approach to assess the risks associated with the emerging development of nickel and cobalt extraction in tropical and subtropical regions of the world (e.g. Reichelt-Brushett and Hudspith 2016).

The use of corals for trace metal studies has been widely documented. A wide range of techniques has been used to extract and measure trace metals in a variety of coral species (Reichelt-Brushett and McOrist 2003), and hence studies are not always directly comparable. However, it becomes necessary to mention about the various studies as they reveal a lot of information regarding emerging pollutants such as heavy metals and their deleterious effects on different coral species in different parts of the world. These studies give an insight regarding metal pollution and their impacts on *scleractinian* corals in various parts of the world.

Some studies have focused on the analysis of the skeleton only (Scott 1990; Guzman and Jimenez 1992; Bastidas and Garcia 1999; Ramos et al. 2004; David 2003; Medina-Elizade et al. 2002; Guzman and Garcia 2002; Livingston and Thompson 1971; and Runnals and Coleman 2003). Whilst for other studies, tissue and skeleton components have been analyzed separately or combined together (Denton and Burdon-Jones 1986; Howard and Brown 1987; Glynn et al. 1989; Harland and Brown 1989; Esslemont 1999; Esslemont et al. 2000; Reichelt-Brushett and McOrist 2003; and Smith et al. 2003). Studies have also focused on the partitioning of metals between tissues and skeleton (McConchie and Harriot 1992), and between tissue, zooxanthellae, gametes and skeleton of selected corals (Reichelt-Brushett and McOrist 2003).

Furthermore, studies have also examined the effects of various metals on the fertilisation success of gametes and include Reichelt-Brushett and Harrison (2005); Reichelt-Brushett and Hudspith (2016); and Gissi et al. (2017). Metal toxicological studies have also been carried out on scleractinian coral larvae (Goh 1991; Reichelt-Brushett and Harrison 2004). These studies show that metals affect all stages of the life cycle of corals.

Synchronous spawning of corals provides a large number of individuals for ecotoxicological testing and as such many studies such as Reichelt-Brushett and Harrison (1999, 2005); Negri and Hoogenboom (2011); Gissi et al. (2017); Leigh-Smith et al. (2018) carried out substantial coral toxicology work on the early coral life stages such as fertilisation success, larvae survival and development. Hudspith et al. (2017) mentioned that fertilisation success is an important endpoint in ecotoxicological assays as it is sensitive and displays a quantifiable dose–response relationship at metal concentrations similar to those polluted environmental levels. However, Summer et al. (2019) reported that trace metal toxicity to early life stages of scleractinian corals varied between both the coral species and metal of interest.

Various reasons have been suggested for the analysis of different components of corals. Coral skeletons are analyzed as they reveal past environmental data of the marine environment. Coral tissues and the associated symbiotic algae tend to record higher metal concentration than the skeleton (Jafarabadi et al. 2018), suggesting that not all metals taken up by the tissue are transferred to the skeleton. Symbiotic zooxanthellae present in tissues of hermatypic corals (Reichelt-Brushett and McOrist 2003) have been described as directly influencing the skeletal concentration of the metals through the enhancement of calcification rates (Livingston and Thompson 1971). Zooxanthellae may be involved in the direct uptake of metals in cases where potentially toxic metals are metabolically substituted for vitally essential elements such as phosphorus (Howard and Brown 1984). Coral gametes and larvae are principally sensitive to changes in water quality because they are released into the environment to fertilise and undergo development where they are potentially in direct contact with anthropogenic contaminants (Reichelt-Brushett and Hudspith 2016).

Stress in corals often gives rise to the expulsion of symbiotic zooxanthellae (Smith et al. 2003), causing pale or white coloration, referred to as bleaching (discussed in Chap 2). Yang et al. (2019) mentioned that coral bleaching is a key reason of decreasing coral reef biodiversity. At higher concentrations, metals can be toxic to corals (Reichelt-Brushett and McOrist 2003; Mohammed and Dar 2010; El-Sorogy et al. 2012). Hence, if corals are exposed to high concentrations of toxic metals, then zooxanthellae can be expelled (Harland and Brown 1989; Harland and Nganro 1990; Meehan and Ostrander 1997; Peters et al. 1997; Reichelt-Brushett and McOrist 2003; Bastidas and Garcia 2004), as they play an important role in the uptake of trace metals.

However, due to different methods and techniques of analysis used, variation of species as well as sample collection from polluted and unpolluted areas, makes it difficult to compare directly the data that have been generated for heavy metals.

3.5 Effects of Heavy Metals in Corals

The recorded metal concentrations in the corals are high and this is generally attributed to the substitution mechanism of the metals into the crystal lattice; the adsorption process; trapping up of the particulate matter into the aragonite lattice and the uptake of organic matter from the coral tissues and feeding through polyps. The free ionic form of metals is the most toxic and bioavailable, and therefore toxicity becomes higher in oligotrophic waters where complexation opportunities are few. Marques et al. (2019) highlighted that metal pollution is a common local scale impact on coral reefs. Nonetheless, sensitivity to trace metals is dependent on a variety of physiological and biochemical factors that are specific to each species as well as external factors such as the duration of exposure (Hudspith et al. 2017). Potential factors such as impaired sperm activation and motility, changes in egg membrane permeability and conductivity, and elevated intracellular reactive oxygen species affect the biochemical and cellular mechanisms of action underlying trace metal toxicity to external fertilisation (which is yet to be fully understood) (review by Hudspith et al. 2017).

Once incorporated within the coral, heavy metals can result in acute or chronic toxicity causing lethal effects or long-term impacts to key biological processes of corals. The key biological processes which are affected include: respiration (Howard et al. 1986); fertilization, metamorphosis and larvae settlement, larvae survival and motility (Goh 1991; Reichelt-Brushett and Harrison 2000; Negri and Heyward 2001; Reichelt-Brushett and Harrison 2004; Reichelt-Brushett and Harrison 2005; Mitchelmore et al. 2007); and complete inhibition of fertilization in *Goniastrea aspera, Favites chinensis* and *Platygyra ryukyuensis* gametes when exposed to copper sulphate solutions greater than or equal to 0.5 mg/L (Heyward 1988). The lethal effects on corals include: reduced growth (Howard and Brown 1987); mucus secretion (Wooldridge 2009); physiological stress (Howard and Brown 1984); loss of zooxanthellae and photosynthetic efficiency in adult corals (Esquivel 1986; Harland and Brown 1989; Harland and Nganro 1990; Jones 1997; Bielmyer et al. 2010; Biscéré et al. 2015); enhanced mortality (Mitchelmore et al. 2007); and reduced biodiversity (Ramos et al. 2004).

Observations by author (s) on the differential effects of heavy metals on corals include: Vaschenko et al. (1999), who mentioned that trace metals also interfere with calcium cation exchange and membrane permeability; Biuki et al. (2010), who mentioned that normal behaviour is affected; Thongra-ar (1997), Vaschenko et al. (1999), and Reichelt- Brushett and Harrison (2005); who stated that heavy metals delay or inhibit normal reproduction and early-life-stage development in many marine invertebrates; Richmond (1997), who stated that exposure to excess trace metal concentrations, interferes with many cellular processes; Thompson et al. (1980) Hughes et al. (2005), and Negri et al. (2005) have observed tissue retraction as a cnidarian stress response.

Stress could also be a behavioral response allowing the detachment of tissue and the escape of coral polyps from the skeleton; van Dam et al. (2011) highlighted about induced bleaching due to higher concentrations of Cd and Cu; Gissi et al. (2017) stated that Cu inhibits coral fertilisation, larval metamorphosis, and survival between 15 and 150 μg Cu/L; Reichelt-Brushett and Harrison (2005), Reichelt-Brushett and Hudspith (2016) and Gissi et al. (2017) found that Ni inhibits fertilisation success at very high concentrations (>1000 μg Ni/L); Goh (1991) stated that 9000 μg Ni/L inhibited larval survival and settlement; Gissi et al. (2019) reported that the toxicity of dissolved Ni on adult corals and their microbiome occurred at 470 μg Ni/L after an exposure of 96 h; Heyward (1988) and Reichelt-Brushett and Harrison (2005) mentioned that high concentrations of Zn had deleterious effects on the fertilization of coral gametes; and Negri et al. (2002) stated that excess Zn concentrations affected the settlement, metamorphosis and survival rate of coral larvae.

Furthermore, Zn can also have deleterious effects on the fertilization of corals gametes (Heyward 1988; Reichelt-Brushett and Harrison 2005) as well as on the settlement, metamorphosis and survival rate of coral larvae (Negri et al. 2002) when present in excess concentrations. In addition to this, Gilbert and Guzman (2001) highlighted that increased Cu exposure in marine environments impair coral physiology through carbonic anhydrase activity. Increased Cu exposure can also lead to declines in photosynthetic efficiency of algal symbionts (Bielmyer et al. 2010). In their study, Reichelt-Brushett and Hudspith (2016) reported that *Platygyra daedalea* gametes validated an acute sensitivity to Cu, having an estimated EC_{50} of 33 μg/L.

Studies concerning the effect of Cu on coral gamete fertilization success include Reichelt-Brushett and Harrison (1999); Negri and Heyward (2001); Heyward (1988). The fertilization success of the gametes was dependent on the Cu concentration. For example: Negri and Heyward (2001) showed that in *Acropora. millepora* < 30% fertilization success occurred at about 70 μg/L Cu; whilst Reichelt-Brushett and Harrison (2005) observed that *Goniastera.aspera* gametes showed 1% fertilization success at 96.8 μg/L Cu.

It is also important to note that seasonal and interannual deviation in Mn concentrations in corals can serve as an indicator of El-Nino Southern Oscillation events and sediment input into coastal waters from rivers (Shen and Sanford 1990; Fallon et al. 2002; Alibert et al.2003; Lewis et al. 2007).

3.6 Heavy Metal Pathways in Corals and Understanding the Role of Zooxanthallae

Heavy metal regulation in marine invertebrates can be acquired through excretion, impaired uptake, detoxification, storage, sequestration of metals, metal-binding to proteins such as glutathione (GSH), metallothioneins (MT) and release of metals via increased production of mucus or nematocyst discharge. GSH functions in metal detoxification via oxidative stress from redox active metals by oxidation of GSH to GSSG (glutathione disulphide); and metals also bind to GSH and be removed from the organism via GSH-conjugation reactions (Mitchelmore et al. 2007).

The speciation of the heavy metals in the marine environment determines their biological availability or potential toxicity of the metals. Factors such as elevated pCO_2 levels and/or low pH in seawater increases the bioavailability of trace elements due to increased solubility and/or changes in the metal speciation. Increased pCO_2 concentrations also affect the uptake/excretion mechanisms of trace elements in marine organisms.

The symbiosis in corals is highly complex with the zooxanthallae contributing to the energy budget of the host and enhancing calcification, thus influencing the skeletal deposition rates of heavy metals. Studies have revealed that high loads of heavy metal concentrations enable the release of symbiotic algae and secretion of mucus in cnidarians. Hence, algae expulsion is used as a mechanism of metal detoxification, mainly because algae have been found to accumulate metals to a larger extent and be more tolerant than their symbiotic hosts. Studies such as Harland et al. (1990); Harland and Nganro (1990) and Reichelt-Brushett and McOrist (2003) have revealed that zooxanthallae play an important role in the uptake of heavy metals. Mucus production (Marshall 2002) and gamete production (Reichelt-Brushett and McOrist 2003) also influence uptake of heavy metals.

Loss of zooxanthallae is a suggested mechanism by which accumulated metal concentrations could be regulated by corals (Reichelt-Brushett and McOrist 2003; Hardefeldt and Reichelt-Brushett 2015). The expulsion of symbiotic zooxanthellae by cnidarians has been reported in response to a number of natural and anthropogenic perturbations including metal exposure (Meehan and Ostrander 1997; Peters et al. 1997; Bastidas and Garcia 2004). The expulsion of metal rich zooxanthallae influences the whole body metal concentrations in corals. However, there is no clear correlation between metal exposure and metal accumulation in corals nor there is for the rate of transfer of metals. Essentially, consideration of uptake and depuration of metals in corals is vital in understanding metal accumulation in corals.

Hardefeldt and Reichelt-Brushett (2015) state that zooxanthellae display similar patterns of metal bioaccumulation to other marine algae, and hence suggest that metal accumulation in zooxanthellae may be a function of uptake and detoxification mechanisms similar to those occurring in other marine algae. The regulation of metals in the living components of corals influences the concentrations of metals transferred to and stored within coral skeleton.

Hardefeldt and Reichelt-Brushett (2015) studied anemone and hence suggest that zooxanthellae exert a strong influence on the whole anemone metal concentrations. Hardefeldt and Reichelt-Brushett (2015) mentioned that variations in zooxanthellae density due to stress and other factors represents an important factor in metal regulation in the living components of cnidarians and must be considered when interpreting skeletal metal concentrations in biomonitoring studies.

Coral bleaching is yet another mechanism through which zooxanthallae is lost and hence metal loads in cnidarians could also be regulated. Coral bleaching transpires when coral-algal symbioses are stressed by high temperatures, high irradiance (both in the visual and ultraviolet regions of the solar spectrum) or other environmental changes, such as manmade pollutants (Bielmyer et al. 2010). Studies such as Harland and Brown (1989); Peters et al. (1997); Reichelt-Brushett and McOrist (2003) and

Shah (2008) have reported about the loss of zooxanthellae in response to metal exposure. These studies give an indication that heavy metals are regulated especially in corals through zooxanthallae loss.

3.7   Summary

Scleractinian corals remain the architects of coral reefs, and consequently a good representative of environmental changes of the ocean. Being sessile, their skeletons integrate records of the presence and the past water chemistry in terms of pollution with respect to concentrations of certain heavy metals, sedimentation rates, sea surface temperature and salinity and ocean acidification. Conducive environmental conditions induce scleractinian corals to incorporate the heavy metals in their living tissue and the skeleton through substitution of dissolved metal species via calcium substitution. Many studies have revealed that tissues and the zooxanthallae tend to record higher concentrations of heavy metals than skeleton.

High metal loads lead to the expulsion of zooxanthallae, causing a pale coloration of the corals, referred to as bleaching of corals. Once incorporated into the coral skeleton; the metals will remain embedded forever since the new growth covers the old carbonate surface and the metals can result in acute or chronic toxicity causing lethal effects or long-term impacts to key biological processes of corals. Ecotoxicological data show the effects of trace metals on various life stages of corals including gamete fertilization; and hence can be used reliably to predict the effects of pollution and thus manage discharges in the marine ecosystems.

Studies also reveal that regulation of heavy metals in the marine invertebrates occurs through processes like excretion, impaired uptake, detoxification, storage, sequestration of metals, metal-binding to proteins such as glutathione (GSH), metallothioneins (MT) and release of metals via increased production of mucus, nematocyst discharge or zooxanthallae loss.

References

Alibert C, Kinsley L, Fallon SJ, McCulloch MT, Berkelmans R, McAllister F (2003) Source of trace element variability in Great Barrier Reef corals affected by the Burdekin flood plumes. Geochim Cosmochim Acta 67(2):231–246. https://doi.org/10.1016/S0016-7037(02)01055-4

Al-Rousan SA, Al-Shloul RN, Al-Horani FA, Abu-Hilal A (2007) Heavy metal contents in growth bands of *Porites* corals: record of anthropogenic and human developments from the Jordanian Gulf of Aqaba. Mar Pollut Bull 54:1912–1922

Babcock RC, Heyward AJ (1986) Larval development of certain gamete-spawning scleractinian corals. Coral Reefs 5:111–116

Baird AH, Cumbo VR, Gudge S, Keith SA, Maynard JA, Tan CH, Woolsey ES (2015) Coral reproduction on world's southernmost reef at Lord Howe Island, Australia. Aquat Biol 23:275–284

Bastidas C, Garcia E (1999) Metal Content on the reef coral *Porites astreoides*: an evaluation of river influence and 35 years of chronology. Mar Pollut Bull 38:899–907

Bastidas C, Garcia EM (2004) Sublethal effects of mercury and its distribution in the coral Porites astreoides. Mar Ecol Prog Ser 267:133–143

Bielmyer GK, Grosell M, Bhagooli R, Bake AC, Langdon C, Gillette PC, TR, (2010) Differential effects of copper on three species of Scleractinian corals and their algal symbionts (*Symbiodinium* spp). Aquat Toxicol 97:125–133

Biscéré T, Rodolfo-Metalpa R, Lorrain A, Chauvaud L, Thébault J, Clavier J, Houlbréque F (2015) Responses of two scleractinian corals to cobalt pollution and ocean acidification. PLoS ONE 10(4):e0122898. https://doi.org/10.1371/journal.pone.0122898

Biuki A, Savari A, Martazavi MS, Zolgharnein H (2010) Acute toxicity of cadmium chloride (CdCl$_2$) on Chanos chanos and their behavioural responses. World J Fish Mar Sci 2(6):481–486

Boboria D, Maata M, Mani FS (2021) Metal pollution in sediments and bivalves in Marovo Lagoon Solomon Islands. Mar Pollut Bull 164:112026. https://doi.org/10.1016/j.marpolbul.2021.112026

Brown BE, Tudhope AW, Le Tissier MDA, Scoffin TP (1991) A novel mechanism for Iron incorporation into coral skeletons. Coral Reefs 10:211–215

Ceh J, Raina JB, Soo RM, van Keulen M, Bourne DG (2012) Coral-bacterial communities before and after a coral mass spawning event on Ningaloo reef. PLoS ONE 7(5):e36920. https://doi.org/10.1371/journal.pone.0036920

Combosch DJ, Vollmer SV (2013) Mixed asexual and sexual reproduction in the Indo-Pacific reef coral Pocillopora damicornis. Ecol Evol 3(10):3379–3387. https://doi.org/10.1002/ece3.721

David CP (2003) Heavy metal concentration in growth bands of corals: a record of mine tailings input through time (Marinduque Islands, Philippines). Mar Pollut Bull 46:187–196

Denton GRW, Burdon-Jones C (1986) Trace metals in corals from the Great Barrier Reef. Mar Pollut Bull 17:209–213

El-Sorogy AS, Mohammed MA, Nour HE (2012) Heavy metals contamination of the Quaternary coral reefs, Red Sea coast Egypt. Environ Earth Sci 67:777–785. https://doi.org/10.1007/s12665-012-1535-0

Esquivel IF (1986) Short term copper bioassay on the planula of the reef coral *Pocillopora damicornis*. In: Jokiel PL, Richmond RH, Rogers RA (eds) Coral reef population biology. Technical Report 37, Hawaii Institute of Marine Biology, Coconut Island, HI, USA, pp 469–472

Esslemont G (1999) Heavy metals in corals from Heron Island and Darwin Harbour Australia. Mar Pollut Bull 389:1051–1054

Esslemont G, Harriot VJ, McConchie DM (2000) Variability of trace-metal concentrations within and between colonies of *Pocillopora damicornis*. Mar Pollut Bull 40:637–642

Fadlallah Y (1983) Sexual reproduction, development and larval biology in scleractinian corals. A Review 2:129–150

Fallon SJ, White JC, McCulloch MT (2002) Porites corals as recorders of mining and environmental impacts: Misima Island, Papua New Guinea. Geochim Cosmochim Acta 66:45–62

Ferrier-Pages C, Houlbreque F, Wyse E, Richard C, Allemand D, Boisson F (2005) Bioaccumulation of zinc in the scleractinian coral *Stylophora pistillata*. Coral Reefs 24:636–645

Gilbert AL, Guzman HM (2001) Bioindication potential of carbonic anhydrase activity in anemones and corals. Mar Pollut Bull 42:742–744

Gissi F, Stauber JL, Reichelt-Brushett A, Harrison PL, Jolley DF (2017) Inhibition in fertilisation of coral gametes following exposure to nickel and copper. Ecotoxicol Environ Saf 145. https://doi.org/10.1016/j.ecoenv.2017.07.009

Gissi F, Reichelt-Brushett AJ, Chariton AA, Stauber JL, Greenfield P, Humphrey C, Salmon M, Stephenson SA, Cresswell T, Jolley DF (2019) The effect of dissolved nickel and copper on the adult coral *Acropora muricata* and its microbiome. Environ Pollut 250:792–806. https://doi.org/10.1016/j.envpol.2019.04.030

Glynn PW, Szmant AM, Corcoran EF, Cofer-Shabica SV (1989) Condition of coral reef Cnidarians from the Northern Florida reef tract: pesticides, heavy metals and histopathological examination. Mar Pollut Bull 20:568–576

Goh BPL (1991) Mortality and settlement success of *Pocillopora damicornis* larvae during recovery from low levels of nickel. Pac Sci 45(3):276–286

Goreau TJ (1977) Coral skeletal chemistry: physiological and environmental regulation of stable isotopes and trace metals in *Montastrea annularis*. Proc R Soc Biol Sci 196:291–315

Guzman HM, Jimenez CE (1992) Contamination of coral reefs by heavy metals along the Caribbean Coast of Central America (Costa Rica and Panama). Mar Pollut Bull 24:554–561

Guzman HM, Garcia EM (2002) Mercury levels in coral reefs along the Caribbean Coast of Central America. Mar Pollut Bull 44:1415–1420

Hardefeldt JM, Reichelt-Brushett AJ (2015) Unravelling the role of zooxanthellae in the uptake and depuration of an essential metal in *Exaiptasia pallida*; an experiment using a model cnidarian. Mar Pollut Bull 96:294–303. https://doi.org/10.1016/j.marpolbul.2015.04.055

Harland AD, Brown BE (1989) Metal tolerance in the *scleractinian* coral *Porites lutea*. Mar Pollut Bull 20:353–357

Harland AD, Nganro NR (1990) Copper uptake by the sea anemone *Anemonia viridis* and the role of zooxanthellae in metal regulation. Mar Biol 104:297–301

Harland A, Bryan G, Brown B (1990) Zinc and cadmium absorption in the symbiotic anemone *Anemonia viridis* and the non-symbiotic anemone *Actiniaequina*. J Mar Biol Assoc U K 70:789–802

Harrison PL (2011) Sexual reproduction of scleractinian corals. In: Dubinsky Z, Stambler N (eds) Coral reefs: an ecosystem in transition. Springer, Netherlands, pp 59–85

Harrison PL, Wallace CC (1990) Reproduction, dispersal and recruitment of scleractinian corals. In: Dubinsky Z (ed) Ecosystems of the world: coral reefs. Elsevier, Amsterdam, pp 133–207

Harrison PL, Ward S (2001) Elevated levels of nitrogen and phosphorus reduce fertilization success of gametes from scleractinian reef corals. Mar Biol 139:1057–1068

Harrison PL, Babcock RC, Bull GD, Oliver JK, Wallace CC, Willis BL (1984) Mass spawning in tropical reef corals. Science 223:1186–1189

Heyward AJ (1988) Inhibitory effects of copper and zinc sulphates on fertilization in corals. In: 6th international coral reef symposium, Australia, pp 299–303

Hidaka M (2016) Life history and stress response of scleractinian corals. In: Kayanne (eds) Coral reef science, coral reefs of the world 5. https://doi.org/10.1007/978-4-431-54364-0_1.

Howard LS, Brown BE (1984) Heavy metals and reef corals. Oceanogr Mar Biol Annu Rev 22:195–210

Howard LS, Brown BE (1987) Metals in *Pocillopora damicornis* exposed to Tin Smelter Effluent. Mar Pollut Bull 18:451–454

Howard LS, Crosby DG, Alino P (1986) Evaluation of some methods for quantitatively assessing the toxicity of heavy metals to corals. In: Jokiel PL, Richmond RH, Rogers RA (eds) Coral reef population biology. Hawaii Institute of Marine Biology. Technical Report No 37

Hudspith M, Reichelt-Brushett AJ, Harrison PL (2017) Factors affecting the toxicity of trace metals to fertilization success in broadcast spawning marine invertebrates: a review. Aquat Toxicol 184:1–13. https://doi.org/10.1016/j.aquatox.2016.12.019

Hughes R, Reichelt-Brushett AJ, Newman LJ (2005) Identifying suitable invertebrate species from a unique habitat for ecotoxicological testing. Australas Soc J Ecotoxicol 11:85–92

Jafarabadi AR, Bakhtiari AR, Maisano M, Pereira P, Cappello T (2018) First record of bioaccumulation and bioconcentration of metals in Scleractinian corals and their algal symbionts from Kharg and Lark coral reefs (Persian Gulf, Iran). Sci Total Environ 640:1500–1511

Jones RJ (1997) Zooxanthallae loss as a bioassay for assessing stress in corals. Mar Ecol Prog Ser 149:163–171

Kerr AM, Baird AH, Hughes TP (2011) Correlated evolution of sex and reproductive mode in corals (Anthozoa: Scleractinia). Proc R Soc B Biol Sci 278:75–81

Khaled A, El Nemr A, El Sikaily A (2003) Contamination of coral reef by trace metal along the Egyptian Red Sea coast. Bull Environ Contam Toxicol 71(3):577–584

Leigh-Smith J, Reichelt-Brushett A, Rose AL (2018) The characterization of Iron (III) in seawater and related toxicity to early life stages of Scleractinian corals. Environ Toxicol 37(4):1104–1114. https://doi.org/10.1002/etc.4043

Lewis SE, Shields GA, Kamber BS, Lough JM (2007) A muti-trace element coral record of land-use changes in the Burdekin River catchment, NE Australia. Palaeogeogr Palaeoclimatol Palaeoecol 246(2):471–487. https://doi.org/10.1016/j.palaeo.2006.10.021

Livingston HD, Thompson G (1971) Trace element concentration in some modern corals. Limnol Oceanogr 16:786–795

Marques JA, Abrantes DP, Marangoni LFB, Bianchini A (2019) Ecotoxicological responses of a reef calcifier exposed to copper, acidification and warming: a multiple biomarker approach. Environ Pollut. https://doi.org/10.1016/j.envpol.2019.113572

Marshall AT (2002) Occurrence, distribution and localization of metals in cnidarians. Microsc Res Tech 56:341–357

McConchie D, Harriot VJ (1992) The partitioning of metals between tissue and skeletal parts of corals: application in pollution monitoring, In: Proceedings of the seventh international coral reef symposium, Guam, 22–26 June, 1992, 97–103

Medina-Elizade M, Gold-Bouchot G, Ceja-Moreno V (2002) Lead contamination in the Mexican Caribbean recorded by the coral *Montastrea annularis* (Ellis Solander). Mar Pollut Bull 44:421–431

Meehan WJ, Ostrander GK (1997) Coral bleaching: a potential biomarker of environmental stress. J Toxicol Environ Health 50:529–552

Mitchelmore CL, Verde EA, Weis VM (2007) Uptake and partitioning of copper and cadmium in the coral *Pocillopora damicornis*. Aquat Toxicol 85:48–56

Mitterer RM (1978) Amino acid composition and metal binding capacity of the skeletal protein of corals. Bull Mar Sci 28:173–180

Mohammed TAAA, Dar MA (2010) Ability of corals to accumulate heavy metals, Northern Red Sea Egypt. Environ Earth Sci 59:1525–1534. https://doi.org/10.1007/s12665-009-0138-x

Negri AP, Heyward AJ (2001) Inhibition of coral fertilization and larval metamorphosis by tributyltin and copper. Mar Environ Res 51(1):17–27. https://doi.org/10.1016/s0141-1136(00)00029-5

Negri AP, Smith LD, Webster NS, Heyward AJ (2002) Understanding ship-grounding impacts on a coral reef: potential effects of anti-foulant paint contamination on coral recruitment. Mar Pollut Bull 44(2):111–117. https://doi.org/10.1016/s0025-326x(01)00128-x

Negri AP, Hoogenboom MO (2011) Water contamination reduces the tolerance of coral larvae to thermal stress. PLoS ONE 6(5):e19703. https://doi.org/10.1371/journal.pone.0019703

Negri AP, Vollhardt C, Humphrey C, Heyward AJ, Jones R, Eaglesham G, Fabricius K (2005) Effects of the herbicide diuron on the early life history stages of coral. Mar Pollut Bull 51:370–383. https://doi.org/10.1016/j.marpolbul.2004.10.053

Peters EC, Gassman NJ, Firman JC, Richmond RH, Power EA (1997) Ecotoxicology of tropical marine ecosystems. Environ Toxicol Chem 16:12–40

Prouty NG, Goodkin NF, Jones R, Lamborg CH, Storlazzi CD, Hughen KA (2013) Environmental assessment of metal exposure to corals living in Castle Harbour, Bermuda. Mar Chem 154:55–66

Rainbow PS (1995) Biomonitoring of heavy metal availability in the marine environment. Mar Pollut Bull 31:183–192

Ramos AA, Inoue Y, Ohde S (2004) Metal contents in *Porites* corals: Anthropogenic input of river run-off into a coral reef from an urbanized area Okinawa. MarPollut Bull 48:281–294

Reichelt-Brushett A (2012) Risk assessment and ecotoxicology: limitations and recommendations for ocean disposal of mine waste in the Coral Triangle. Oceanography 25(4):40–51. https://doi.org/10.5670/oceanog.2012.66

Reichelt-Brushett AJ, Harrison PL (1999) The effect of copper, zinc and cadmium on fertilization success of gametes from scleractinian reef corals. Mar Pollut Bull 38(3):182–187

Reichelt-Brushett AJ, Harrison PL (2000) The effect of copper on the settlement success larvae from the *scleractinian* coral *Acropora tenuis*. Mar Pollut Bull 41:385–391

Reichelt-Brushett, AJ, McOrist G (2003) Trace metals in the living and nonliving components of scleractinian corals. Mar Pollut Bull 46:1573–1582

Reichelt-Brushett AJ, Harrison PL (2004) Development of a sublethal test to determine the effects of copper and lead on scleractinian coral larvae. Arch Environ Contam Toxicol 47:40–55. https://doi.org/10.1007/s00244-004-3080-7

Reichelt-Brushett AJ, Harrison PL (2005) The effect of selected trace metals on the fertilisation success of several scleractinian coral species. Coral Reefs 24:524–534. https://doi.org/10.1007/s00338-005-0013-5

Reichelt-Brushett AJ, Michalek-Wagner K (2005) Effects of copper on the fertilisation success of the soft coral *Lobophytum compactum*. Aquat Toxicol 74:280–284

Reichelt-Brushett AJ, Hudspith M (2016) The effects of metals of emerging concern on the fertilization success of gametes of the tropical scleractinian coral *Platygyra daedalea*. Chemosphere 150:398–406

Richmond RH (1997) Reproduction and recruitment in corals: critical links in the persistence of reefs. In: Birkeland C (ed) Life and death of coral reefs. Chapman & Hall, New York, pp 175–197

Rotchell JM, Ostrander GK (2011) Molecular toxicology of corals: a review. J Toxicol Environ Health Part B 14:571–592

Runnalls LA, Coleman ML (2003) Record of natural and anthropogenic changes in reef environments (Barbados West Indies) using laser ablation ICP-MS and sclerochronology on coral cores. Coral Reefs 22:416–426

Schyff V, Yive NSCK, Bouwman H (2020) Metal concentrations in corals from South Africa and the Mascarene Basin: a first assessment for the Western Indian Ocean. Chemosphere 239:124784. https://doi.org/10.1016/j.chemosphere.2019.124784

Scott PJB (1990) Chronic pollution in coral skeletons in Hong Kong. J Exp Mar Biol Ecol 139:51–64

Shah S (2008) Study of heavy metal accumulation in *scleractinian* corals of Viti Levu, Fiji Islands, unpublished MSc thesis. Faculty of Science, Technology and Environment, The University of the South Pacific, Fiji.

Shen GT, Boyle EA (1988) Determination of lead, cadmium and other trace metals in annually-banded corals. Chem Geol 67:47–62

Shen GT, Sanford CL (1990) Trace element indicators of climate variability in reef building corals. In: Glynn (PW) Global ecological consequences of the 1982–83 El Nino Southern Oscillation. Elsevier, Amsterdam, pp 255–283. https://doi.org/10.1016/S0422-9894()8)70038-2.

Smith LD, Negri AP, Philipp E, Webster NS, Heyward AJ (2003) The effects of antifoulant-paint-contaminated sediments on coral recruits and branchlets. Mar Biol 143:651–657

St John BE (1974) Heavy metals in the skeletal carbonate of *scleractinian* corals. In: Proceedings of the second international coral reef symposium, Brisbane, pp 461–469.

Summer K, Reichelt-Brushett A, Howe P (2019) Toxicity of manganese to various lfe stages of selected marine cnidarian species. Ecotoxicol Environ Saf 167:83–94. https://doi.org/10.1016/j.ecoenv.2018.09.116

Thompson JH, Shinn EA, Bright TJ (1980) Chapter16: Effects of drilling muds on seven species of reef-building corals as measured in the field and laboratory. Oceanography Series 27 (Part A):433–453

Thongra-ar W (1997) Toxicity of cadmium, zinc and copper on sperm cell fertilization of sea urchin, Diadema setosum. J Sci Soc Thail 23:297–306

van Dam RA, Harford AJ, Houstan MA, Hogan AC, Negri AP (2008) Tropical marine toxicity testing in Australia: a review and recommendations. Australas J Ecotoxicol 14:55–88

van Dam JW, Negri AP, Uthicke S, Mueller JF (2011) Chemical pollution on coral reefs: exposure and ecological effects. In: Sández-Bayo F, van den Brink PJ, Mann RM (eds) Ecological impacts of toxic chemicals. Francisco Sanchez-Bayo, Sydney, pp 187–211

Vaschenko MA, Zhanga ZP, Lama PKS, Wua RSS(1999) Toxic effects of cadmium on fertilizing capability of spermatozoa, dynamics of the first cleavage and pluteus formation in the sea urchin *Anthocidaris crassispina* (Agassiz). Mar Pollut Bull 38(12):1097–1104

Ward S, Harrison P (2000) Changes in gametogenesis and fecundity of acroporid corals that were exposed to elevated nitrogen and phosphorous during ENCORE experiment. J Exp Biol Ecol 246:179–221

Ward S, Harrison P, Hoegh-Guldberg O (2002) Coral bleaching reduces reproduction of scleractinian corals and increases susceptibility to future stress. In: Proceedings of the ninth international coral reef symposium 2 Bali, October 2000, pp 1123–1128

Wooldridge SA (2009) A new conceptual model for the enhanced release of mucus in symbiotic reef corals during "'bleaching'" conditions. Mar Ecol Prog Ser 396:145–152

Yang T, Chen H, Wang H, Drews M, Sennan Li, Huang W, Zhou H, Chen CM, Diao X (2019) Comparative study of polycyclic aromatic hydrocarbons (PAHs) and heavy metals (HMs) in corals, surrounding sediments and surface water at the Dazhou Island, China. Chemosphere 218:157–168. https://doi.org/10.1016/j.chemosphere.2018.11.063

Chapter 4
Analytical Techniques

Abstract This chapter tabulates the analytical techniques used for selected studies and briefs about the species of *scleractinian* coral used; the component of the coral analysed; the appropriate analytical methodology adopted; the different metals analysed and the specific instrument used for analysis. Different concentrations of heavy metals are obtained for different studies due to different components being analysed and the concentrations measured as either wet weight or dry weight. Certain studies focus on the analysis of the total metal concentrations for the skeleton, tissue and zooxanthallae whilst some studies have analysed specific components of the coral species. Hence, this chapter gives a general feedback on the analytical techniques adopted for the variety of *scleractinian* coral species under study. This sort of feedback can be used for general comparison between studies and *scleractinian* species and the different analytical techniques.

Keywords Analytical techniques · Total metal concentration · *Scleractinian* corals

4.1 Analytical Techniques Used for Heavy Metal Analysis in Various *Scleractinian* Coral Species

Various analytical techniques have been used to analyze heavy metal concentrations in different components of diverse corals. Heavy metal analysis was carried out either on the skeleton or the tissue, or both the skeleton and tissue combined together or even on the skeleton, tissue, zooxanthellae and gametes. The following table provides a brief summary of the analytical techniques used for selected studies mentioned in this monograph.

© The Author(s), under exclusive license to Springer Nature Switzerland AG 2021   71
S. B. Shah, *Heavy Metals in Scleractinian Corals*, SpringerBriefs in Earth Sciences,
https://doi.org/10.1007/978-3-030-73613-2_4

References	Species used	Analytical methodology used	Components Analysed	Metals analysed	Instrument used for analysis
Goreau (1977)	*Montastrea annularis*	HCl	Skeleton	Mg Sr Fe Al	Atomic Absorption Spectrophotometry (AAS)
Denton and Burdon-Jones (1986)	*Acropora formosa* *Fungia concinna Verrill* *Fungia Fungites*	Nitric acid/Perchloric acid digestion Refluxing 4 days Digestate solvent extracted	Skeleton and tissue	Cu Zn Cd Pb Ni	Atomic Absorption Spectroscopy (AAS)
Howard and Brown (1987)	*Pocillopora damicornis*	Tissues removed and acid digested using 30:1 nitric acid and perchloric acid Skeleton acid digested	Tissues Skeleton	Cu Zn Fe	Inductively Coupled Plasma Spectrophotometer (ICP)
Harland and Brown (1989)	*Porites lutea*	Tissues removed and acid digested using 30:1 nitric acid and perchloric acid	Tissues	Fe	Atomic Absorption Spectroscopy (AAS)
Scott (1990)	*Platygyra sinensis*	Skeleton samples were dissolved in dilute HCl and analysed	Skeleton	Al Cd Cu Pb V U Y	Inductively Coupled Plasma Mass Spectroscopy (ICPMS)

(continued)

(continued)

References	Species used	Analytical methodology used	Components Analysed	Metals analysed	Instrument used for analysis
McConchie and Harriott (1992)	*Acropora aspera* *Acropora valida* *Pocillopora damicornis* *Goniastrea favulus* *Porites spp*	Tissue digested using hydrogen peroxide Skeleton digested using Nitric acid	Skeleton Tissue	Zn Cd Pb Cu	Anodic Stripping Voltammetry (ASV)
Bastidas and Garcia (1999)	*Porites astreoides*	Nitric acid/peroxide digestion	Skeleton	Cd Ca Cu Fe Pb Zn Cr	Inductively Coupled Plasma-Atomic Emission Spectroscopy (ICP-AES)
Esslemont (1999)	*Goniastrea aspetra* *Acropora nobilis* *Montastrea annuligera*	Removal of organic and non-coralline components	Skeleton	Cd Cu Pb	Anodic Stripping Voltammetry (ASV)
		Acid digestion at room temperature using HCl and lanthanum chloride		Zn Ni Cr	Atomic Absorption Spectroscopy (AAS)
Esslemont et al. (2000)	*Pocillopora damicornis*	Acid digestion of tissues and skeleton	Tissues Skeleton	Zn Ni Cr Cd Pb	Atomic Absorption Spectroscopy (AAS)
				Cu	Anodic Stripping Voltammetry (ASV)

(continued)

(continued)

References	Species used	Analytical methodology used	Components Analysed	Metals analysed	Instrument used for analysis
Medina-Elizade et al. (2002)	*Montastraea annularis*	Samples dissolved in nitric acid and chelated using APDC/MIBK	Skeleton	Pb	Atomic Absorption Spectroscopy (AAS)
Mokhtar et al. (2002)	*Porites spp* *Favia spp*	Nitric acid H2O2: NaOH (1:1)	Coral skeletons	Cr Mn Hg Zn	Neutron Activation Analysis (NAA)
Reichelt-Brushett and McOrist (2003)	*Acropora tenius*	Separation of tissue and zooxanthellae from skeleton Acid digestion of all components using 1:1 nitric acid and peroxide on a hot plate for 3hrs	Zooxanthellae Tissue Skeleton	Pb Cd Zn Cu Fe Ni Mn As	Neutron Activation Analysis (NAA) and Inductively Coupled Plasma Mass Spectroscopy (ICPMS)
Smith et al.(2003)	*Acropora formosa* *Acropora microphthalma*	Butyltins extracted using methanol/HCl Volatile tin hydrides formed and packed in columns Cu/Zn-acid digested	Coral branches were sediment treated	TBT(Sn) DBT(Sn) Mbt(Sn)	Graphite Furnace Atomic Absorption Spectrophotometry (GFAAS)
				Cu Zn	Inductively Coupled Plasma-Atomic Emission Spectroscopy (ICP-AES)
Bastidas and Garcia (2004)	*Porites astreoides*	Acid digestion for 5 h in a water bath	Polyp, zooxanthellae and skeleton	Hg	Cold Vapor Atomic Absorption Spectroscopy (CVAAS)

(continued)

(continued)

References	Species used	Analytical methodology used	Components Analysed	Metals analysed	Instrument used for analysis
Ramos et al.(2004)	*Porites*	Nitric acid digestion	Skeleton	Na Mg Al Ca Mn Zn As Sr Y Ag Cd Ba Pb Th U	Inductively Coupled Plasma Mass Spectroscopy (ICPMS)
Reichelt-Brushett and Harrison (2004)	*Goniastera aspera*	Acidification	Water samples	Cu Pb	Graphite Furnace Atomic Absorption Spectrophotometry (GFAAS)

(continued)

(continued)

References	Species used	Analytical methodology used	Components Analysed	Metals analysed	Instrument used for analysis
Madkour (2005)	*Stylophora wellesi, Acropora humilis, Acropora hemprichii, Acropora hyacinthus, Pocillopora damicornis, Millepora complanata and Millepora dichotoma Porites lutea, and Porites compressa*	Acid digestion in a Teflon cup using a mixture of conc. nitric, perchloric and hydrofluoric acids, (3: 2: 1) respectively	Whole sample	Fe Mn Zn Cu Pb Ni Co Cd	Atomic Absorption Spectroscopy (AAS)
Reichelt-Brushett and Harrison (2005)	*Acropora tenuis Acropora longicyathus Goniastrea aspera Goniastrea retiformis*	Acid not specified	Sperm-free seawater	Cu Pb Zn Cd Ni	Graphite Furnace Atomic Absorption Spectrophotometry (GFAAS)
Al-Rousan et al. (2007)	*Porites*	X-ray/EPA #3050..acid digestion of nitric acid and hydrogen peroxide	Skeleton	Cd Cu Fe Mn Zn Pb	Flame Atomic Absorption Spectroscopy (FAAS)
Mitchelmore et al. (2007)	*Pocillopora damicornis*	Nitric acid and hydrogen peroxide	Coral homogenate and tissues	Cu Cd	Inductively Coupled Plasma Mass Spectroscopy (ICPMS)

(continued)

(continued)

References	Species used	Analytical methodology used	Components Analysed	Metals analysed	Instrument used for analysis
Dar et al. (2008)	*Echinopora Genmacea,Iichinopora genmaceu, Lopophyllia coiymbosa, Gulaxiufiusicularis, Porites lutea* and *Fubliu* sp.	Concentrated nitric acid digestion	Bleached and non-bleached specimen of the samples	Fe Zn Cu Pb Co	Atomic Absorption Spectroscopy (AAS)
Prouty et al. (2008)	*Montastraea faveolata*	Acid leaching, oxidising, reducing using Miili-Q water; nitric acid	Skeleton; annual growth band	Al B Ba Cd Co Cr Cu Fe Mg Mn Ni Pb Sb Sn Sr U V Zn Ca	High-Resolution Double Focusing Magnetic Sector-Field Inductively Coupled Plasma Mass Spectrometer (HR-SF-ICP-MS)

(continued)

(continued)

References	Species used	Analytical methodology used	Components Analysed	Metals analysed	Instrument used for analysis
Shah(2008)	*Acropora Formosa* *Pocillopra damicornis* *Porites spp*	1:1 nitric acid and hydrogen peroxide	Zooxanthellae, tissue and skeleton	Cu Pb Cd Zn Fe	Atomic Absorption Spectroscopy (AAS)
Sabdono (2009)	*Galaxea fascicularis*	Acid digestion with HF, HNO_3, $HClO_4$	Coral sample	Cu	Atomic Absorption Spectrophotometer (AAS)
Bielmyer et al.(2010)	*A. cervicornis, M. faveolata, P. damicornis*	Soft tissue fractions digested with 1 N nitric acid	coral and zooxanthallae	Cu	Graphite Furnace Atomic Absorption Spectrophotometry (GFAAS)
Chen et al. (2010)	*Porites*	Nitric acid digestion	Skeleton	Fe Mn Zn Sr Ca Mg	Inductively Coupled Plasma-Atomic Emission Spectroscopy (ICP-AES)

(continued)

(continued)

References	Species used	Analytical methodology used	Components Analysed	Metals analysed	Instrument used for analysis
Mohammed and Dar (2010)	*Acropora cythearea, A. humilis, A. hyacnthus, Echinoporagemmacea, Favialaxa, Pocillopora verrucosa, P. damicornis, Porites solida, Stylophora pistillata, Millepora complanata, and Favia sp.*	Nitric acid digestion	Coral sample	Fe Cu Pb Ni Mn Zn	Atomic Absorption Spectrophotometer (AAS)
Prouty et al. (2010)	*Porites lobata*	Outer tissues removed from skeleton using a water pik	Skeleton	^{43}Ca ^{84}Sr ^{89}Y ^{138}Ba ^{139}La ^{140}Ce ^{238}U ^{57}Fe ^{27}Al	Laser Ablation Inductively Coupled Mass Spectrometry (LA-ICP-MS)

(continued)

(continued)

References	Species used	Analytical methodology used	Components Analysed	Metals analysed	Instrument used for analysis
Ali et al. (2011)	*Acropora hemprichi* *Acropora pharaonis* *Acropora humilis* *Stylophora pistillata*	Hot nitric acid	Skeleton and tissue	Cu Zn Pb Cd Ni Co Fe	Atomic Absorption Spectrophotometry (AAS)
El-Sorogy et al. (2012)	*Stylophora pistillata;* *Pocillopora verrucosa;* *Acropora cytherea;* *Porites lutea;* *Fungiafungites;* *Goniastrea pectinata;* *Favites* *Pentagona*	Acid digested in 5 mL of HCl and 15 mL HNO_3–$HClO_4$ (5:1)	Coral sample	Zn Pb Mn Fe Cr Co Ni Cu	Inductively Coupled Plasma-Atomic Emission Spectroscopy (ICP-AES)
Mokhtar et al. (2012)	*Hydnophora microconos;* *Favia speciose;* *Porites lobata*	Skeleton: nitric acid digestion after cleaning; sediment: HCl: HNO_3 digestion	Coral skeleton, sediment and seawater samples	Cu Zn Ni Cd Fe Mn	Flame Atomic Absorption Spectrophotometry (FAAS)
Berry et al. (2013)	*Porites furcate* *Agariciatenuifolia*	Treatment with nitric acid, perchloric acid and hydrochloric acid	Tissue and sediment	As Cd Cu Zn	Inductively Coupled Plasma Optical Emission Spectrometry (ICP-OES)

(continued)

(continued)

References	Species used	Analytical methodology used	Components Analysed	Metals analysed	Instrument used for analysis
Prouty et al. (2013)	Diplorialabyrinthiformis	Acid digestion using ultra-pure concentrated nitric acid	Coral powder	Pb Zn Mn Ca Hg	High-Resolution Double Focusing Magnetic Sector-Field Inductively Coupled Plasma Mass Spectrometer (HR-SF-ICP-MS)
Barakat et al. (2014)	Stylophora sp, Pocillopora sp, Acropora sp, Fungia sp, Porites sp	Skeleton: 1:1 nitric acid: water, nitric acid, hydrogen peroxide, HCl Tissue: 1:1 nitric acid: hydrogen peroxide, HCl, lanthanum chloride	Skeleton and tissue separately	Cd Cu Pb Zn Ni Cr	Flame Atomic Absorption Spectroscopy (FAAS)
Chan et al. (2014)	Tubastraea coccinea	Nitric acid digestion	Skeleton and tissue separately	Al Cd Co Cr Cu Fe Mn Ni Pb Zn Ca	Inductively Coupled Plasma Optical Emission Spectrometer (ICP-OES)

(continued)

(continued)

References	Species used	Analytical methodology used	Components Analysed	Metals analysed	Instrument used for analysis
Song et al. (2014)	*Porites*	Nitric acid	Skeleton	Cr Mn Ni Cu Zn Cd Ba Pb Sr U Ca	Inductively Coupled Plasma Mass Spectrometer (ICP-MS)
Chen et al. (2015)	*Porites*	Nitric acid digestion	Skeleton	Mn Cu V	Inductively Coupled Plasma Mass Spectrometer (ICP-MS)
Darvishnia et al. (2016)	*Faviidae and Portiidae species*	Nitric acid: perchloric acid (4:1)	Skeleton	Pb Fe Zn	Flame Atomic Absorption Spectroscopy (FAAS) and Graphite Furnace Atomic Absorption Spectrophotometry (GFAAS)
Reichelt-Brushett and Hudspith (2016)	*Platygyra daedalea*	Acid not specified	Sperm-free seawater	Ni Co	Inducticely Coupled Plasma Mass Spectrometry (ICPMS)
Gissi et al. (2017)	*Platygyra daedalea Acropora aspera Acropora digitifera*	Nitric acid	Coral gametes	Ag Al, As, Ba, Cd, Co, Cr, Cu, Fe, Mn, Ni, Pb, Se, V, Zn	Inductively Coupled Plasma-Atomic Emission Spectroscopy (ICP-AES)

(continued)

(continued)

References	Species used	Analytical methodology used	Components Analysed	Metals analysed	Instrument used for analysis
Bielmyer-Fraser et al. (2018)	*Acropora cervicornis* *Pocillopora damicornis*	Nitric acid	Tissue fractions (coral and zooxanthallae)	Cu	Atomic Absorption Spectroscopy (AAS)
Deschaseaux et al. (2018)	*Acropora aspera*	1: 3 nitric acid: hydrochloric acid	Tissue and zooxanthallae fraction	Zn	Inductively Coupled Plasma Mass Spectrometer (ICP-MS)
Summer et al. (2019)	*Acropora spathulata* *Platygyra daedalea*	Nitric acid	Adult coral and gametes	Mn	Inductively-Coupled Plasma Mass-Spectrometry (ICP-MS)
Yang et al. (2019)	*Acropora hyacinthus*	Nitric acid: hydrochloric acid (3:1) microwave acid dissolution method	Adult coral ~ 5–6 cm snips	Cr Mn Ni Cu Zn As Cd Pb	Inductively-Coupled Plasma Mass Spectrometry (ICP-MS)
Gillmore et al. (2020)	*Acropora muricata*	Pre-digestion using 2:1 nitric acid: hydrogen peroxide; Microwave heating	Algal symbionts and coral tissue	Ni	Inductively Coupled Plasma-Atomic Emission Spectroscopy (ICP-AES)

(continued)

(continued)

References	Species used	Analytical methodology used	Components Analysed	Metals analysed	Instrument used for analysis
Li et al. (2020)	*Acropora nana* *Golden calcareous coral* *Galaxea Fascicularis* *Sinularia sp* *Favites abdita* *Montipora danae* *Favia speciose* *Pavona decussate*	Acid microwave digestion: 6 ml of 65% HNO_3; 2 ml of 35% H_2O_2	Coral species	V Cr Mn Fe Co Ni Cu Zn As Se Mo Cd Pb U	Inductively Coupled Plasma Mass Spectrometry (ICP-MS)

(continued)

(continued)

References	Species used	Analytical methodology used	Components Analysed	Metals analysed	Instrument used for analysis
van der Schyff et al. (2020)	*Coral fragments*	Acid digestion using: 33% hydrochloric acid: 50% hydrogen peroxide: 65% conc. Nitric acid	Coral fragments (all)	Al Ti V Cr Mn Fe Co Ni Cu Zn As Cd Hg Pb U Se	Inductively Coupled Plasma Mass Spectrometry (ICPMS)
Nakamura et al. (2020)	*Porites Lutea*	Nitric acid digestion	100 mg of coral powder from coral slabs	Fe Zn Ni Cu Cr Mn Pb	Inductively Coupled Plasma Mass Spectrometry (ICPMS)

4.2 Summary

Extensive studies have been carried out on the study of heavy metals as marine pollutants and their effects on *scleractinian* corals. A variety of scleractinian corals have been utilized for heavy metal analysis for the above mentioned studies. The coral components which have been analysed include the skeleton separately, skeleton and tissue combined together as well as separately, zooxanthallae of the coral species, gametes of corals species as well as the whole coral sample. The analytical methodology adopted by majority of the studies was acid digestion. However, the combination of the acids and the digestion time varied amongst the studies. Use of instrumentation amongst the studies also varied and hence concentration levels (either as parts per million or parts per billion) of metals in studies depended on the sensitivity of the instruments. Therefore, studies were not directly comparable in terms of concentration levels of the heavy metals studied. However, this chapter gives an impartial awareness about the methodology to be used in for the various coral species and the instruments to be used for analysis.

References

Ali AAM, Hamed MA, Abd El-Azim H (2011) Heavy metals distribution in the coral reef ecosystems of the Northern Red Sea. Helgol Mar Res 65:67–80. https://doi.org/10.1007/s10152-010-0202-7

Al-Rousan SA, Al-Shloul RN, Al-Horani FA, Abu-Hilal A (2007) Heavy metal contents in growth bands of *Porites* corals: record of anthropogenic and human developments from the Jordanian Gulf of Aqaba. Mar Pollut Bull 54:1912–1922

Barakat SA, Al-Rousan S, Al-Trabeen MS (2014) Possible index for marine pollution of scleractinian corals in Northern Gulf of Aqaba, Jordan. J Nat Sci Res 4(18). ISSN 2225-0921

Bastidas C, Garcia E (1999) Metal content on the reef coral *Porites astreoides*: an evaluation of river influence and 35 years of chronology. Mar Pollut Bull 38:899–907

Bastidas C, Garcia EM (2004) Sublethal effects of mercury and its distribution in the coral *Porites astreoides*. Mar Ecol Prog Ser 267:133–143

Berry KLE, Seemann J, Dellwig O, Struck U, Wild C, Leinfelder RR (2013) Sources and spatial distribution of heavy metals in scleractinian coral tissues and sediments from the Bocas Del Toro Archipelago Panama. Environ Monit Assess 185:9089–9099. https://doi.org/10.1007/s10661-013-3238-8

Bielmyer GK, Grosell M, Bhagoolic R, Baker AC, Langdon C, Gillette P, Capo TR (2010) Differential effects of copper on three species of *scleractinian* corals and their algal symbionts (Symbiodinium spp.). Aquat Toxicol 97:125–133

Bielmyer-Fraser GK, Patel P, Capo T, Grosell M (2018) Physiological responses of corals to ocean acidification and copper exposure. Mar Pollut Bull 133:781–790

Chan I, Hung JJ, Peng SH, Tseng LC, Ho TY, Hwang JS (2014) Comparison of metal accumulation in azooxanthellate coral (*Tubastraea coccinea*) from different polluted environments. Mar Pollut Bull 85:648–658. https://doi.org/10.1016/j.marpolbul.2013.11.015

Chen T, Yu Ke-Fu Li S, Price GJ, Qi S, Wei G (2010) Heavy metal pollution recorded in Porites corals from Daya Bay, Northern South China Sea. Mar Environ Res 70(3–4):318. Elsevier. https://doi.org/10.1016/j.marenvres.2010.06.004.hal-00613293

Chen X, Wei G, Deng W, Liu Y, Sun Y, Zeng T, Xie L (2015) Decadal variations in trace metal concentrations on a coral reef: evidence from a 159-year record of Mn, Cu, and V in a Porites coral from the northern South China Sea. J Geophys Res Oceans 120:405–416. https://doi.org/10.1002/2014JC010390

Dar M, Ali AEA, Murad FA (2008) Response of *scleractinian* corals to the natural and anthropogenic heavy metal stresses in the Northern Red Sea and Gulfs of Suez and Aqaba. Egypt J Aquat Res 34(4):126–142

Darvishnia Z, Bakhtiari AR, Kamrani E, Sajjadi MM, Fakharzadeh M (2016) Biomonitoring of Pb, Fe and Zn using corals as bioindicator species, Qeshm Island, the Persian Gulf. Indian J Geo Mar Sci 45(12):1704–1708

Denton GRW, Burdon-Jones C (1986) Trace metals in corals from the Great Barrier Reef. Mar Pollut Bull 17:209–213

Deschaseaux E, Hardefeldt J, Jones G, Reichelt-Brushett A (2018) High zinc exposure leads to reduced dimethylsulfoniopropionate (DMSP) levels in both the host and endosymbionts of the reef-building coral Acropora aspera. Mar Pollut Bull 126:93–100. https://doi.org/10.1016/j.marpolbul.2017.10.070

El-Sorogy AS, Mohammed MA, Nour HE (2012) Heavy metals contamination of the Quaternary coral reefs, Red Sea coast Egypt. Environ Earth Sci 67:777–785. https://doi.org/10.1007/s12665-012-1535-0

Esslemont G (1999) Heavy metals in corals from Heron Island and Darwin Harbour Australia. Mar Pollut Bull 389:1051–1054

Esslemont G, Harriot VJ, McConchie DM (2000) Variability of trace-metal concentrations within and between colonies of *Pocillopora damicornis*. Mar Pollut Bull 40:637–642

Gillmore ML, Gissi F, Golding LA, Stauber JL, Reichelt-Brushett AJ, Severati A, Humphrey CA, Jolley DF (2020) Effects of dissolved nickel and nickel-contaminated suspended sediment on the scleractinian coral *Acropora Muricata*. Mar Pollut Bull 152:110886. https://doi.org/10.1016/j.marpolbul.2020.110886

Gissi F, Stauber J, Reichelt-Brushett A, Harrison PL, Jolley DF (2017) Inhibition in fertilisation of coral gametes following exposure to nickel and copper. Ecotoxicol Environ Saf 145:32–41. https://doi.org/10.1016/j.ecoenv.2017.07.009

Goreau TJ (1977) Coral skeletal chemistry: physiological and Environmental regulation of stable isotopes and trace metals in *Montastrea annularis*. Proc R Soc Lond Series B Biol Sci 196(1124):291–315. https://www.jstor.org/stable/77080

Harland AD, Brown BE (1989) Metal tolerance in the *scleractinian* coral *Porites lutea*. Mar Pollut Bull 20:353–357

Howard LS, Brown BE (1987) Metals in *Pocillopora damicornis* exposed to Tin Smelter Effluent. Mar Pollut Bull 18:451–454

Li J, Sun C, Zhang L, Ding J, Jiang F, Wang Z, Wang Z, Fu L (2020) Current distribution characteristics of trace elements in the coral-reef systems of Xisha Islands China. Mar Pollut Bull 150:110737. https://doi.org/10.1016/j.marpolbul.2019.110737

Madkour HA (2005) Geochemical and environmental studies of recent marine sediments and some hard corals of Wadi El-Gemal area of the red sea *Egypt*. Egypt J Aquat Res 31(1)

McConchie D, Harriot VJ (1992) The partitioning of metals between tissue and skeletal parts of corals: application in pollution monitoring. In: Proceedings of the seventh international coral reef symposium, Guam, 22–26 June, pp 97–103

Medina-Elizade M, Gold-Bouchot G, Ceja-Moreno V (2002) Lead contamination in the Mexican Caribbean recorded by the coral *Montastrea annularis* (Ellis Solander). Mar Pollut Bull 44:421–431

Mitchelmore CL, Verde EA, Weis VM (2007) Uptake and partitioning of copper and cadmium in the coral *Pocillopora damicornis*. Aquat Toxicol 85:48–56

Mohammed TAAA, Dar MA (2010) Ability of corals to accumulate heavy metals, Northern Red Sea Egypt. Environ Earth Sci 59:1525–1534. https://doi.org/10.1007/s12665-009-0138-x

Mokhtar MB, Praveena SM, Aris AZ, Yong OC, Lim AP (2012) Trace metal (Cd, Cu, Fe, Mn, Ni and Zn) accumulation in Scleractinian corals: a record for Sabah, Borneo. Mar Pollut Bull 64:2556–2563

Mokhtar MB, Wood AKBH, Hou-Weng C, Ling TM, Sinniah AA (2002) Trace metals in selected corals of Malaysia. Online J Biol Sci 2(12):805–809

Nakamura N, Kayanne H, Takahashi Y, Sunamura M, Hosoi G, Yamano H (2020) Anthropogenic anoxic history of the Tuvalu Atoll recorded as annual black bands in coral. Sci Rep 10:7338. https://doi.org/10.1038/s41598-020-63578-4

Prouty NG, Hughen KA, Carilli J (2008) Geochemical signature of land-based activities in Caribbean coral surface samples. Coral Reefs. https://doi.org/10.1007/s00338-008-0413-4

Prouty NG, Field ME, Stock JD, Jupiter SD, McCulloch M (2010) Coral Ba/Ca records of sediment input to the fringing reef of the Southshore of Moloka'i, Hawai'i over the last several decades. Mar Pollut Bull 60:1822–1835

Prouty NG, Goodkin NF, Jones R, Lamborg CH, Storlazzi CD, Hughen KA (2013) Environmental assessment of metal exposure to corals living in Castle Harbour Bermuda. Mar Chem 154:55–66

Ramos AA, Inoue Y, Ohde S (2004) Metal contents in *Porites* corals: Anthropogenic input of river run-off into a coral reef from an urbanized area Okinawa. Mar Pollut Bull 48:281–294

Reichelt-Brushett AJ, McOrist G (2003) Trace metals in the living and nonliving components of scleractinian corals. Mar Pollut Bull 46:1573–1582

Reichelt-Brushett AJ, Harrison PL (2004) Development of a sublethal test to determine the effects of copper and lead on scleractinian coral larvae. Arch Environ Contam Toxicol 47:40–55

Reichelt-Brushett AJ, Harrison PL (2005) The effect of selected trace metals on the fertilization success of several scleractinian coral species. Coral Reefs 24:524–534. https://doi.org/10.1007/s00338-005-0013-5

Reichelt-Brushett AJ, Hudspith M (2016) The effects of metals of emerging concern on the fertilization success of gametes of the tropical scleractinian coral *Platygyra daedalea*. Chemosphere 150:398–406

Sabdono A (2009) Heavy metal levels and their potential toxic effect on coral *Galaxea fascicularis* from Java Sea Indonesia. Res J Environ Sci 3(1):96–102

Scott PJB (1990) Chronic pollution in coral skeletons in Hong Kong. J Exp Mar Biol Ecol 139:51–64

Shah SB (2008) Study of heavy metal accumulation in *scleractinian* corals of Viti Levu, Fiji Islands, unpublished MSc thesis. Faculty of Science, Technology and Environment, The University of the South Pacific, Fiji.

Smith LD, Negri AP, Philipp E, Webster NS, Heyward AJ (2003) The effects of antifoulant- paint-contaminated sediments on coral recruits and branchlets. Mar Biol 143:651–657

Song Y, Yu K, Zhao J, Feng Y, Shi Q, Zhang H, Ayoko GA, Frost RL (2014) Past 140-year environmental record in the northern South China Sea: evidence from coral skeletal trace metal variations. Environ Pollut 185:97–106. https://doi.org/10.1016/j.envpol.2013.10.024

Summer K, Reichelt-Brushett A, Howe P (2019) Toxicity of manganese to various life stages of selected marine cnidarian species. Ecotoxicol Environ Saf 167:83–94. https://doi.org/10.1016/j.ecoenv.2018.09.116

Van der Schyff V, du Preez M, Blom K, Kylin H, Yive NSCK, Merven J, Raffin J, Bouwman H (2020) Impacts of a shallow shipwreck on a coral reef: a case study from St. Brandon's Atoll, Mauritius, Indian Ocean. Mar Environ Res 156:104916. https://doi.org/10.1016/j.marenvres.2020.104916

Yang T, Chen H, Wang H, Drews M, Sennan Li, Huang W, Zhou H, Chen CM, Diao X (2019) Comparative study of polycyclic aromatic hydrocarbons (PAHs) and heavy metals (HMs) in corals, surrounding sediments and surface water at the Dazhou Island, China. Chemosphere 218:157–168. https://doi.org/10.1016/j.chemosphere.2018.11.063

Conclusion/Outlook for the Future

This monograph has given an insight on heavy metals; a marine pollutant and their effects on *scleractinian* corals, which are reef building corals. Heavy metals are perpetual additions to the marine environment and are toxic, persistent and non-biodegradable and do not get degraded by bacteria. Entry to the marine environment is by natural as well as anthropogenic means. Being sessile in nature, corals accumulate these metals and hence represent good environmental indicators.

This monograph has studied the current state of knowledge on the status quo of heavy metals in the marine environment and its effects of the *scleractinian* corals; heavy metal regulation in corals and the possible analytical techniques through which these metals can be measured. A holistic view of heavy metals and their effects on the corals has been deliberated.

Recommendations for future include investigating the route of uptake and transfer of the heavy metals and the mechanism of metal detoxification and regulation in corals for the various coral species. Since chemical speciation is a common phenomenon in the marine environment, detailed information on this with respect to the different heavy metals and their respective effects on corals could be looked into. As new studies emerge regarding heavy metals as eminent pollutants in the marine environment, their effects on the various coral species could be studied in detail.

Toxicological studies of heavy metal pollution in marine environments need to take into account the interactions amongst metals, which may influence the uptake, accumulation and toxicity of various metals synchronously. Toxicological and ecotoxicological studies should aim to provide relevant information on heavy metal pollution from anthropogenic sources, such that these do not give rise to adverse effects on marine organisms such as corals. In addition, ecotoxicological studies need to consider the differences in the bioavailability of heavy metals amongst different species of corals. An in-depth knowledge of the escalating pollution in combination with other anthropogenically—induced environmental changes and their consequences on the global shifts in climate and oceanographic conditions on heavy metal toxicity is needed. Despite the spatial and temporal distributions of tropical keystone species; namely *scleractinian corals*, a work in progress needs to be articulated to discuss ways to destress the already stressed coral reefs.

© The Author(s), under exclusive license to Springer Nature Switzerland AG 2021   89
S. B. Shah, *Heavy Metals in Scleractinian Corals*, SpringerBriefs in Earth Sciences,
https://doi.org/10.1007/978-3-030-73613-2

The practicality of DMSP and DMS and their possible antioxidant roles in corals could also be emphasized on. It would be ideal to study the relationship of the rate of influx of DMS and the fact that it helps in regulating climate. Furthermore the effects of various heavy metals on DMSP could also be looked into.

Printed in the United States
by Baker & Taylor Publisher Services